Thomas Bewick, James Edmund Harting

Our Summer Migrants

An Account of the Migratory Birds Which Pass the Summer in the British Islands

Thomas Bewick, James Edmund Harting

Our Summer Migrants
An Account of the Migratory Birds Which Pass the Summer in the British Islands

ISBN/EAN: 9783337813017

Printed in Europe, USA, Canada, Australia, Japan

Cover: Foto ©berggeist007 / pixelio.de

More available books at **www.hansebooks.com**

OUR SUMMER MIGRANTS.

AN ACCOUNT OF

THE MIGRATORY BIRDS

WHICH PASS THE SUMMER IN

THE BRITISH ISLANDS.

BY J. E. HARTING, F.L.S., F.Z.S.

AUTHOR OF A "HANDBOOK OF BRITISH BIRDS,"
A NEW EDITION OF WHITE'S "SELBORNE,"
ETC., ETC.

ILLUSTRATED FROM DESIGNS BY THOMAS BEWICK.

LONDON:
BICKERS AND SON,
1, LEICESTER SQUARE.
1875.

PREFACE.

FOR those who reside in the country and have both leisure and inclination to observe the movements and habits of birds, there is not a more entertaining occupation than that of noting the earliest arrival of the migratory species, the haunts which they select, and the wonderful diversity which they exhibit in their actions, nidification, and song.

There is something almost mysterious in the way in which numbers of these small and delicately formed birds are found scattered in one day over a parish where on the previous day not one was to be seen; and the manner of their arrival is scarcely more remarkable than the regularity with which they annually make their appearance.

That most of them reach this country after long and protracted flights, crossing the Mediterranean, the Bay of Biscay, and the English Channel is an undoubted fact. They have been seen to arrive upon our shores, and have been observed at sea during their passage, often at a considerable distance from land.

But how few of those who notice them in this country know where they come from, why they come, what they find here to live upon, how, when, and where they go for the winter!

In the following chapters an attempt has been made to answer these questions, and to give such information generally about our summer migratory birds as will prove acceptable to many who may be glad to possess it without knowing exactly where to look for it. Some of these sketches were originally published in the Natural History columns of "The Field" during the summer of 1871, and as a reprint has frequently been asked for, I have now carefully revised them and made some important additions and emendations, besides adding to the series a dozen or more chapters which have never before appeared.

The illustrations, from designs by Thomas Bewick, will, it is conceived, add considerably to the attractiveness of the volume, and will enable the reader to dispense with particular descriptions of the species, which it might be otherwise desirable to furnish. These may be found, moreover, in other works devoted to British Ornithology.

JAMES EDMUND HARTING.

July, 1875.

CONTENTS.

	Page
THE Wheatear	1
The Whinchat	9
The Stonechat	13
The Wood Warbler	16
The Willow Warbler	24
The Chiff-chaff	28
The Nightingale	32
The Blackcap	44
The Orphean Warbler	51
The Garden Warbler	59
The Common Whitethroat	67
The Lesser Whitethroat	71
The Redstart	74
The Sedge Warbler	81
The Reed Warbler	83
The Grasshopper Warbler	86
Savi's Warbler	88
The Aquatic Warbler	91
The Marsh Warbler	92
The Great Reed Warbler	101
The Rufous Warbler	103
The Pied Wagtail	106
The White Wagtail	110
The Grey Wagtail	112
The Yellow Wagtail	117

	Page
The Grey-headed Wagtail	121
The Meadow Pipit	124
The Rock Pipit	130
The Tree Pipit	135
The Water Pipit	138
Richard's Pipit	142
The Tawny Pipit	146
The Pennsylvanian Pipit	149
The Red-throated Pipit	152
The Spotted Flycatcher	155
The Pied Flycatcher	160
The Swallow	170
The Martin	184
The Sand Martin	187
The Common Swift	191
The Alpine Swift	199
The Nightjar	204
The Cuckoo	219
The Wryneck	242
The Hoopoe	249
The Golden Oriole	262
The Red-backed Shrike	276
The Turtle-dove	282
The Landrail or Corncrake	288
General Observations	299
Conclusion	330
Index	335

THE WHEATEAR.

(*Saxicola œnanthe.*)

ONE of the earliest of our feathered visitors to arrive is the Wheatear, which comes to us as a rule in the second week of March; and, although individuals have been seen and procured occasionally at a much earlier date, there is reason to believe that the spring migration does not set in before this, and that the birds met with previously are such as have wintered in this country; for it has been well ascertained that the Wheatear, like the

Stonechat, occasionally remains with us throughout the year. It is a noticeable fact that those which stay the winter are far less shy in their habits, and will suffer a much nearer approach.

The name Wheatear may have been derived either from the season of its arrival, or from its being taken in great numbers for the table at wheat harvest. Or, again, it may be a corruption of *whitear*, from the *white ear* which is very conspicuous in the spring plumage of this bird. Many instances are on record of Wheatears having come on board vessels several miles from land at the period of migration, and from the observations of naturalists in various parts of the country it would appear that these birds travel by night, or at early dawn. I do not remember any recorded instance in which they have been seen to land upon our shores in the daytime.

In Ireland, according to Mr. Thompson,[1]

---

[1] " Nat. Hist. Ireland ;" Birds, i. pp. 176, 177.

the Wheatear arrives much later than in England, and does not stay the winter. With regard to Scotland, Macgillivray states[1] that it is nowhere more plentiful than in the outer Hebrides, and in the Orkney and Shetland Islands; and from the fact of his having observed the species near Edinburgh on the 28th of February, we may infer that a few, as in England, occasionally remain throughout the year.

The number of Wheatears which used to be taken years ago upon the South Downs in autumn was a matter of notoriety.

"Hereabouts," says an old chronicle of Eastbourne, "is the chief place for catching the delicious birds called Wheatears, which much resemble the French Ortolans;" and Wheatears play an important part in the history of this town. Squire William Wilson, of Hitching, Lord of the Manor of East-Bourne, was in Oliver Cromwell's time vehemently suspected

---

[1] "Hist. Brit. Birds," ii. p. 292.

of loyalty to the Stuarts; and one Lieutenant Hopkins, with a troop of dragoons, swooped down on Eastbourne to search the squire's house, and, if needful, arrest him as a Malignant. The squire was laid up with the gout; but Mistress Wilson, his true wife, with the rarely-failing shrewdness of her sex, placed before Lieutenant Hopkins and his troopers a prodigious pie filled with Wheatears, "which rare repast," the chronicle goes on to say, "the soldiers did taste with so much amazement, delight, and jollity," that the squire upstairs had ample time to burn all the papers which would compromise him, and when Lieutenant Hopkins, full of Wheatear pie, came to search the house, there was not so much treasonable matter found as could have brought a mouse within peril of a *præmunire*. At the Restoration the Lord of the Manor became Sir William Wilson of Eastbourne, a dignity well earned by his devotion to the Royal cause; but the chronicle goes on to hint that Charles II. was passionately fond of Wheatears, and that possibly the liberality of the squire, in supplying his Majesty's table with these delicacies, may have had something to do with the creation of the baronetcy.

The abundance of Wheatears at certain seasons on the Hampshire downs was noticed by Gilbert White in a letter to the Hon. Daines Barrington in Dec. 1773. Since this excellent naturalist penned his observations, however, many changes in the haunts and habits of birds have been remarked. For example, the Hawfinch, which he referred to as "rarely seen in England, and only in winter," is now found to be resident throughout the year, and nesting even in the proximity of London and other large cities. The Landrail, which he noted as "a bird so rare in this district that we seldom see more than one or two in a season, and those only in autumn," is now so plentiful in the same neighbourhood that I have shot as many as half a dozen in one day in September, within a few miles of Selborne. The Common Bunting, which in 1768 was considered to be a "rare bird" in the district referred to, may now be heard there in full song—if song it can be called—throughout the month of May. Whilst walking from Liss to Selborne, I have on two occasions met with a bird which Gilbert White had not observed—the Cirl Bunting; and, to return to the Wheatears, these birds, which were formerly so plentiful in autumn that the shepherds trapped them by dozens, are now far less numerous at the same season, and the practice of snaring them has perceptibly declined.[1] It was remarkable that, although in the height of the season—*i.e.*, at wheat harvest—so many hundreds of dozens were taken, yet they were never seen to flock, and it was a rare thing to see more than three or four at a time; so that there must have been a perpetual flitting and constant progressive succession.

The Wheatear is partial to commons and

---

[1] As to other changes in the fauna and flora which have taken place since Gilbert White's day in the district of which he wrote, the reader may be referred to the Preface to my edition of the "Natural History of Selborne" recently published.

waste lands, old quarries, sand hills, and downs by the sea, and it is in these situations that we may now look for him without much fear of disappointment. Like all the chats, the Wheatear is very terrestrial in its habits, seldom perching on trees, although often to be seen on gate-posts and rails, where a broader footing is afforded it. Its song is rather sprightly, and is occasionally uttered on the wing. The contrast between the spring and autumn plumage of this bird is very remarkable. If an old bird be examined in September, it will be found that the white superciliary streak has almost disappeared; the colour of the upper parts has become reddish brown; the throat and breast pale ferruginous, lighter on the flanks and belly; while the primaries and tail at its extremity are much browner. On raising the feathers of the back, it will be found that the base of each feather is grey; and in spring this colour supersedes the brown of winter, which is worn off, and the upper parts assume a beautiful bluish grey, while the under parts become pure white. In this species, therefore, it is evident that the seasonal change of plumage is effected by a change of colour in the same feather, and not by a moult.

The nest of the Wheatear is generally well concealed in the crevice of a cliff or sandbank, or in an old rabbit burrow. Where these conveniences are not accessible, the nest may be found at the foot of a bush, screened from view by grass or foliage. The eggs, five or six in number, are of a delicate pale blue, occasionally spotted at the larger end with pale rust colour.

The geographical range of the Wheatear is very extensive for so small and short-winged a bird. It is found in the Faroe Isles, Iceland, and Greenland; in Lapland, Norway, Sweden, and Denmark; throughout Europe to the Mediterranean; in Egypt, Arabia, Asia Minor, and Armenia.

THE WHINCHAT.

(Saxicola rubetra.)

SELDOM appearing before the end of the first week in April, the Whinchat arrives much later than the Wheatear, and is much less diffused than that species. By the end of September it has again left the country, and I have never met with an instance of its remaining in England during the winter months. On several occasions correspondents have forwarded to me in winter a bird which they believed to be the Whinchat, but which invariably proved to be a female, or male in winter plumage, of the Stonechat—a species which is known to reside with us throughout the year, yet receiving a large accession to its numbers in spring, and undergoing corresponding decrease in autumn.

In the southern counties of England the Whinchat is sometimes very numerous, and may be found in every meadow perched upon the tall grass stems or dockweed. The abundance or scarcity of this species, however, varies considerably according to season. In some years I have noticed extraordinary numbers of this little bird, and in others have scarcely been able to count two or three pairs in a parish. I have generally found that a cold or wet spring has so affected their migration as to cause them apparently to alter their plans, and induce them to spend the summer but a short distance to the north or north-west of their winter quarters.

It is a little remarkable that in Ireland the Whinchat is far less common than the Stonechat, the reverse being the case in England.

Mr. Thompson says, in the work already quoted (p. 175), " In no part of Ireland have I seen the Whinchat numerous, and compared with the Stonechat it is very scarce." In the south of Scotland, according to Macgillivray, it seldom makes its appearance before the end of April, that is, more than a fortnight after its arrival in England. It extends to Sutherland, Caithness, and the outer Hebrides (*cf.* More, " Ibis," 1865, p. 22), and has occasionally been met with in Orkney, but not in Shetland. In winter it migrates to the south-east, and at that season is not uncommon in Egypt, Nubia, and Abyssinia, travelling also through Asia Minor, Arabia and Persia, as far eastward as the north-west provinces of India. In a south-westerly direction this species, passing through Spain and Portugal, proceeds down the west coast of Africa to Senegal, Gambia, and Fantee.

The Whinchat differs a good deal in its habits from the Wheatear, and on this account, as well as on account of certain differences of structure, it has been placed with the Stonechat and other allied species in a separate genus (*Pratincola*). It is doubtful, however, whether these differences are sufficient to entitle them to anything more than a specific separation.

The Whinchat perches much more than does the Wheatear, and may be seen darting into the air for insects, after the manner of a Flycatcher. It derives its name, of course, from the fact of its being found upon the *whin*, or *furze*, a favourite perch also for its congener the Stonechat. The derivation of the word *whin* I have never been able to ascertain.

Although the two species are frequently confounded, the Whinchat may be always distinguished from the Stonechat by its superciliary white streak, by the lighter-coloured throat and vent, and by the white bases of the three outer tail feathers on each side. Both species make a very similar nest, which is placed on the ground and well concealed, and lay very similar eggs, of a bright blue faintly speckled at the large end with rust colour.

THE STONECHAT.

(Saxicola rubicola.)

AS has been already stated, the Stonechat may be found in a few scattered pairs throughout the country all the year round. At the beginning of April, however, a considerable accession to its numbers is observed to take place, owing to a migration from the south and south-east. It takes up its residence on moors and heaths, and many a lonely walk over such ground is enlivened by the sprightly actions and sharp "*chook-chook*" of this little bird.

The male in his wedding dress, with jet black head, white collar, and ferruginous breast, is extremely handsome; and the artist who is fond of depicting bird-life would scarcely find a prettier subject than a male Stonechat in this plumage upon a spray of furze in full bloom.

In Ireland the Stonechat is considered to be a resident species, and this is attributed by Mr. Thompson to the mild winters of that island. In Scotland, on the contrary, Sir Wm. Jardine has observed that the Stonechat is not nearly so abundant as either the Whinchat or the Wheatear, and frequents localities of a more wild and secluded character. It ranges, however, to the extreme north of the mainland of Scotland, and is included by Dr. Dewar in his list of birds which he found nesting in the Hebrides. It is said not to breed in either Orkney or Shetland.[1]

The geographical range of the Stonechat is rather more extensive than that of the Whin-

---

[1] *Cf.* More, "Ibis," 1865, p. 22.

chat, for besides being found throughout the greater part of Europe to the Mediterranean, it goes by way of Senegal to South Africa, and extends eastward through Asia Minor, Palestine, and Persia, to India and Japan. In Europe, however, its distribution is somewhat remarkable, inasmuch as it is confined chiefly to the central and southern portions of the continent, and in Norway and Sweden is unknown. The Whinchat, on the other hand, breeds in these countries, and has been met with as far north as Archangel. In winter the male Stonechat loses the black head, and the colours in both sexes are much less vivid than in summer. Here again, as with the Wheatear, the change of plumage seems to be effected by a change of colour in the same feathers, and not by a moult.

Apropos of this subject, the reader may be referred to an article contributed by me to the Natural History columns of " The Field," 16th September, 1871, on variation of colour in birds.

THE WOOD WARBLER.

(Phylloscopus sibilatrix.)

ALTHOUGH often taken to comprehend every species of warbler, Professor Newton has recently shown[1] that the genus *Sylvia* of Latham should be restricted to the group of fruit-eating warblers next to be described, and that the generic term which has priority for the willow wren group is *Phylloscopus* of Boie.

From its larger size, brighter colour, and finer song, the Wood Warbler deserves to be first

---

[1] Yarrell, "Hist. Brit. Birds," 4th ed. vol. i. pp. 427, 442.

noticed; and the first step should be to distinguish it from its congeners. Perhaps none of the small insectivorous birds have been more confounded one with another than have the members of this group, not only by observers of the living birds, but by naturalists with skins of each before them. Taking the three species which annually visit us—*i.e.*, the Wood Warbler, the Willow Warbler, and the Chiff-chaff—it will be found on comparison that they differ in size as follows—

	Length.	Wing.	Tarsus.
Wood Warbler	5·2 in.	3·0 in.	0·7 in.
Willow Warbler	5·0 ,,	2·6 ,,	0·7 ,,
Chiff-chaff	4·7 ,,	2·4 ,,	0·6 ,,

Not only is the Wood Warbler the largest of the three, but it has comparatively the longest wings and the longest legs. The wings, when closed, cover three-fourths of the tail. In the Willow Wren, under the same circumstances, less than half the tail is hidden. The Chiffchaff's wing is shorter again. In my edition of White's "Selborne," founded upon that of Bennett, 1875, pp. 56, 57, will be found a long footnote on the subject, with woodcuts illustrating the comparative form of the wing in these three birds. Mr. Blake-Knox, in "The Zoologist" for 1866, p. 300, has pointed to the second quill-feather, depicted in a sketch accompanying his communication, as being an unfailing mark of distinction.[1] When we reflect, however, upon the variation which is found to exist in the length of feathers, owing to the age of the bird, moult, or accident, too much stress ought not to be laid upon this as a character. At the same time there is no doubt that, taken in connection with other details, it will often assist the determination of a species. After examining a large series of these birds, I have come to the conclusion that, as regards the wings, the following formulæ may be relied on: Wood Warbler, 2nd=4th; 3rd and 4th with

---

[1] Mr. Blake-Knox subsequently corrected his statement, remarking that he had by mistake written *second* instead of *third* primary quill. The first primary is so rudimentary as almost to escape observation.

outer webs sloped off towards the extremity. Willow Warbler, 2nd=6th ; 3rd, 4th, and 5th sloped off. Chiff-chaff, 2nd=7th ; 3rd, 4th, 5th, and 6th sloped off.

The Wood Warbler is much greener on the back and whiter on the under parts than either of its congeners, and has a well-defined superciliary streak of sulphur-yellow, which, in the Willow Wren, is much shorter and paler. The legs of the Wood Warbler and Willow Wren are brownish flesh-colour, while those of the Chiff-chaff are dark brown. After the first moult, the young of all three species are much yellower in colour than their parents. Hence the mistake which Vieillot made in describing the young of *P. trochilus* as a distinct species under the name of *flaviventris*.

Although the majority of the *Sylviidæ* are fruit-eaters, the species now under consideration are almost entirely insectivorous ;[1] they are also more strictly arboreal in their habits,

---

[1] Dr. Bree states that he has occasionally observed the Willow Wren taking currants from his trees.

and as regards the character of their nests, they differr emarkably from other members of the *Sylviidæ* in building domed nests on or near the ground, instead of cup-shaped nests at a distance from it. The Yellow-billed Chiff-chaff—or Icterine Warbler, as it should now be called [1]—however, forms an exception to the rule, as will be seen later. As these little birds make their appearance at a season when caterpillars and destructive larvæ begin to be troublesome, the good they do in ridding the young leaves and buds of these pests is incalculable. I have watched a Willow Wren picking the green *aphis* off a standard rose-tree, and have been as much astonished at the quantity which it consumed as at the rapidity of the consumption. The Wood Warbler is not nearly so sociable as either the Willow Warbler or the Chiff-chaff. It keeps to the tops of trees in woods and plantations, and seldom comes into gardens; hence it is not so often seen. Al-

---

[1] See Professor Newton's edition of Yarrell's "History of British Birds," vol. i. p. 360.

though not rare, it is somewhat local, and in the British Islands, it appears, is confined exclusively to England and the south of Scotland. Mr. Thompson has included it with hesitation amongst the birds of Ireland; for although the description given to him of certain birds and eggs seemed to apply to this species, it was stated that the nest which contained the eggs was lined with feathers. Now, the Willow Wren invariably makes use of feathers for this purpose, but the Wood Warbler does not. The nest of the latter is composed entirely of dry grass and leaves, occasionally mixed with a little moss; and although I have sometimes found horsehair inside, I do not remember to have seen or heard of an instance in which any feathers were employed. The eggs, five or six in number, are white, closely freckled over with reddish brown.

Mr. Blake-Knox, a well-known naturalist, resident in the county of Dublin, says ("Zoologist," 1866, p. 300), "I tried very hard this year to add the Wood Wren to our Dublin avifauna, and though I killed some dozens of snowy-white-bellied Willow Wrens, they were all the common *Sylvia trochilus*. That the bird is Irish I am sure, for I have heard it. Should an Irish ornithologist see this, will he try for it, if he should live in a wooded district, such as the counties Wicklow and Wexford? I am sure it is neglected for want of a certain distinction." Since this note was published, the Wood Wren has actually been obtained in Ireland, a specimen having been shot in the county of Fermanagh by Sir Victor Brooke, and preserved by him in June, 1870. Another was obtained the same year at Glen Druid in the county of Dublin, as reported by Mr. Blake-Knox. Both Sir William Jardine and Macgillivray have referred to the Wood Warbler being found northward to the middle districts of Scotland, a circumstance which appears to have been overlooked by Mr. Yarrell, since he says (vol. i. p. 349, 3rd edit.), "I am not aware of any record of its appearance in Scotland." This statement, however, has been rectified in the fourth edition of this standard work by Professor Newton, who remarks: "In Scotland it is known to breed regularly in the counties of Dumfries, Wigton, Lanark and Berwick, the Lothians and Perthshire, and occasionally in those of Roxburgh, Selkirk, Renfrew and Stirling." Mr. A. G. More, in an article "On the Distribution of Birds in Great Britain during the Nesting Season," published in the "Ibis" for 1865, observes (p. 26), that the Wood Warbler "in Scotland ranges further north than the Chiff-chaff, having been observed by the Duke of Argyle in Argyleshire and at Balmoral."

According to Mr. Robert Gray, of Glasgow, it has been observed in Inverness and Aberdeenshire, and Mr. Edwards has found it in Banffshire.

Beyond the British Islands the Wood Warbler is found throughout Europe, though rare in the north, and it extends eastward to Siberia and southward to Algeria, Egypt and Abyssinia. It arrives in this country generally about the middle of April, and leaves again in September.

THE WILLOW WARBLER.

(*Phylloscopus trochilus.*)

THE Willow Warbler is much more generally distributed than the last-named bird; but it is possible that it is considered commoner from the difference in the haunts of the two species—the Wood Warbler, as already remarked, keeping further away from habitations. As a rule, the Willow Wren arrives in this country about the end of the first week in April—that is to say, before the Wood Warbler, but not so early as the Chiff-chaff, which is the first of the genus to appear.

Yarrell speaks of these birds as " having acquired with us the general name of Willow Warblers, or Willow Wrens, from their prevailing green colour;" but Thompson, in his " Birds of Ireland " (i. p. 192), says, "this name was doubtless bestowed upon the bird originally on account of its partiality to willows, which I have frequently remarked, the twigs and branches of the common osier (*Salix viminalis*) abounding with *aphides*, being on such occasions its chief favourite." There is yet another suggestion—*i.e.*, that the name may have been bestowed from the circumstance that these little birds make their appearance just as the willow is budding.

It is marvellous how these tiny creatures can sustain the protracted flights which are necessary to transport them from their winter to their summer quarters ; and yet that they make these long journeys is well ascertained. On the 23rd of April a Willow Wren came on board a vessel eighty miles from Malta and fifty from Cape Passaro, the nearest land. Two days later another alighted on the rigging sixty miles from Calabria, and one hundred and thirty-five from Mount Etna. On the 26th of April, eighty miles from Zante and one hundred and thirty from Navarino, a Willow Wren and a Chiff-chaff were found dead on board, presumably from exhaustion, as they were apparently uninjured. Many other such instances are on record.

The present species may be regarded as the commonest of the three which visit us, being generally dispersed in favourable localities over the whole of Great Britain and Ireland. Although it has not been met with in the Hebrides, the Willow Wren has occasionally been seen in Orkney, and the late Dr. Saxby has recorded a single instance of its occurrence in Shetland. Through every country in Europe it seems to be well known as a periodical migrant.

The winter quarters of the Willow Wren are to a certain extent those of its congeners, that is to say, Northern Africa and Palestine, where it is very numerous in the cold season, but it has been found much further southward. Mr. Ayres sent a specimen to Mr. Gurney from Natal; the late Mr. Andersson met with it in Damaraland, S.W. Africa; and Mr. Layard some years since procured specimens at the Cape. As is often the case with allied species, the remarks as to habits and food which have been applied to the Wood Warbler will apply almost equally well to the present species. The distinction between the birds themselves has been already pointed out. The nests of the Willow Wren and Chiff-chaff are both lined with feathers, the eggs of the former being white spotted with red; while those of the latter are white spotted with purple, chiefly at the larger end.

Varieties in this group of birds are rarely met with, and it may therefore be worth notice that in May, 1861, a primrose-coloured Willow Wren was shot at Witley Park, in the parish of Witley, Surrey, and forwarded for inspection to the editor of " The Field."

THE CHIFF-CHAFF.

(*Phylloscopus rufa.*)

ALTHOUGH the smallest of the three species, the Chiff-chaff is apparently the hardiest of them all, for it often braves the winds of March, and makes its appearance in England long before the leaves have given signs of approaching summer. As I have already pointed out the means of distinguishing this little bird from its congeners, and have referred to its nest and eggs, it will suffice to state that, like the Willow Wren, it is a regular summer visitant to England, Scotland, and Ireland; that it is the earliest of the summer warblers to visit us; and that it remains with us until the first week of September, when it migrates to the south-east to spend the winter in a warmer climate. It appears to be common at that season in Italy, Sicily, the Maltese Islands, and Asia Minor; and Mr. Blyth has found it as far to the eastward as Calcutta.

Old English authors, who knew the Garden Warbler as the Greater Pettychaps, gave the Chiff-chaff the name of the Lesser Pettychaps, presumably from its general resemblance to it in miniature. These two names, however, may now be considered as obsolete.

Whilst on the subject of Willow Warblers, we may refer to the fact that a single example of another species, *P. hypolais* (*vel icterina*, the oldest name for it), which is common enough on the other side of the German Ocean, is recorded to have been taken in England, and another in Ireland. The bird is known as the Yellow-billed Chiff-chaff, Melodious Willow Warbler, and

Icterine Warbler.[1] So long ago as June, 1848, the English specimen referred to was killed at Eythorne, near Dover, and the fact was communicated by Dr. Plomley to Mr. Yarrell, who published it in his "History of British Birds." A second British example of this species was shot at Dunsinea, county Dublin, in June, 1856, and is now in the Royal Dublin Society's Museum.[2] In size it equals the Wood Warbler, and resembles it somewhat in colour, but it has a shorter wing (2·75 in. instead of 3 in.); the whole of the under parts are sulphur-yellow, and the legs and toes are slate colour. These characters may serve to distinguish it at once should it again be met with by ornithologists in England. Should its song be heard, all doubts would at once be set at rest, for as a warbler it is far superior to any of the three species just mentioned. I have had many opportunities

---

[1] See Professor Newton's edition of Yarrell's "History of British Birds," vol. i. p. 360.

[2] This specimen was recorded at the time by Dr. Carte in the "Journal of the Royal Dublin Society," vol. i. p. 440.

of seeing and hearing this little bird in Holland, and can testify to the power and variety of its song. Frequently I contrived to get within a few feet of it, and could almost see the notes as they poured out of its tiny throat. The eggs when fresh are the most lovely imaginable, being of a bright pink with dark purple spots, scattered chiefly at the larger end. The nest, as I have already hinted, is cup-shaped, and placed at a little height from the ground; the bird in this respect departing from the usual habit of the Willow Warblers.

These notes being intended rather as suggestions for those who desire to know a little about our summer birds, than as a condensed history of the species, I may observe, in concluding this chapter, that those who are anxious to glean further particulars about the Willow Warblers and their allies, will do well to consult an excellent article on the subject by Professor Schlegel, published (in French) in 1851 in the " Proceedings of the Royal Zoological Society of Amsterdam."

THE NIGHTINGALE.

(Philomela luscinia.)

IN common with one or two allied species, the Nightingale differs so materially in structure and habits from the garden or fruit-eating warblers (*Sylvia*), with which it has been generally associated, that most naturalists nowadays are agreed in regarding it as the type of a separate genus (*Philomela*). For want of a better English name, and as indicating their haunt, the members of this genus may be called "thicket warblers." As regards structure, they differ from the Garden Warblers in having the bill less compressed towards the tip, and wider near the gape; the legs much longer and not scutellated, the toes more adapted for walking than perching. In habits they are more retired, concealing themselves in thickets and copses, living a good deal on the ground, where they find the principal portion of their food, and building a loosely-constructed nest on or near the ground, instead of a more compact structure at a distance from it.

The sole representative of this genus in England is the far-famed Nightingale; and of all the summer migrants to this country, no species probably has attracted more attention, or given rise to more speculation and discussion amongst naturalists. The most remarkable fact in connection with its annual sojourn in England is its very partial distribution. When we find this bird in summer as far to the westward as Spain and Portugal, and as far to the northward as Sweden, we may well be surprised at its absence from Wales, Ireland,

and Scotland; and yet it is the fact that the boundary line, over which it seldom if ever flies, excludes it from Cornwall, West Devon; part of Somerset, Gloucester, and Hereford; the whole of Wales (*à fortiori* from Ireland), part of Shropshire, the whole of Cheshire, Westmoreland, Cumberland, Durham, and Northumberland. I am well aware that the Nightingale has been stated to have been heard and seen in Wales, Cumberland, and even in Mid-Lothian (see "Zoologist," p. 241); but, even if they could be relied on in every case, which is doubtful, these instances can only be regarded as exceptional. In those counties only to the east of the line indicated can the bird be considered a regular summer visitant. Mr. Blyth has expressed the opinion[1] that the Nightingale migrates almost due north and south, deviating but a very little indeed either to the right or left. "There are none in Brittany," he says, "nor in the Channel Islands,

---

[1] Note to his edition of White's "Selborne," 1836, p. 141.

and the most westward of them probably cross the Channel at Cape la Hogue, arriving on the coast of Dorsetshire, and thence apparently proceeding northwards, rather than dispersing towards the west; so that they are only known as accidental stragglers a little beyond the third degree of western longitude." They arrive generally about the end of the second week in April, and it is a well-ascertained fact that the males invariably precede the females by several days. In 1867 three London birdcatchers, between April 13 and May 2, took 225 Nightingales, and the whole of these, with five or six exceptions only, were cock birds. The previous year these same bird-catchers had supplied the dealer by whom they were employed with 280 Nightingales, of which not more than sixty were hens. From these statistics we may infer that in no locality would Nightingales be more plentiful if unmolested than in the neighbourhood of London; but if one dealer only is instrumental in capturing between 200 and 300 in the season, it is easy to account for the scarcity of the species. On the arrival of the hen birds the cocks soon pair, and assist in building, during which time, and during the time the hens are sitting, they are in full song. When the young are hatched the males leave off singing, and busy themselves in bringing food to the nest.

The song generally ceases before the end of the first week in June. Occasionally, however, I have heard a Nightingale sing on throughout June, but accounted for this by supposing that the nest had been robbed, and that the cock was singing while the hen hatched a second brood. Naturalists who live in London need not travel more than five miles from Charing Cross to hear the Nightingale in full song. Nay, a friend who is well acquainted with the note, has heard the bird frequently in Victoria Park, which is only two miles distant from the Bank of England, and on several occasions attentive observers have recognized the unmistakable notes of the Nightingale in the Botanical Gardens, Regent's Park, and in Kensington Gardens.

It is curious how wide-spread is the belief that the Nightingale warbles only at eve. The reason, no doubt, is that amidst the general chorus by day its song is less noticed or attended to. But that it sings constantly by day is a fact, of which we have satisfied ourselves repeatedly. Moreover, it is by no means the only bird to sing at night. The Sedge Warbler, Grasshopper Warbler, Woodlark, Skylark, and Thrush, may often be heard long after sunset; while the Cuckoo is frequently to be heard at midnight, and the Landrail constantly.

It would appear that of the large number of persons who profess a love for song birds very few, comparatively, have the ear to distinguish a song unless they can see the author of it. Hence it frequently happens that they listen to a Thrush or Blackcap in the early spring, and immediately inform their friends that they have heard the Nightingale weeks before it has reached this country.

Many poets have perpetuated the odd belief

that the mournful notes of the Nightingale are caused by the bird's leaning against a thorn to sing! Shakespeare, for example, in his " Passionate Pilgrim," says:

> " Everything did banish moan,
> Save the nightingale alone.
> She, poor bird, as all forlorn,
> Lean'd her breast up-till a thorn;
> And there sung the dolefull'st ditty,
> That to hear it was great pity."

These lines, by the way, although generally attributed to Shakespeare, and included in most editions of his poems, were written, it is said, by Richard Barnefield in 1598, and published by him in a work entitled "Poems in divers humors."[1] Shakespeare's Lucrece, however, invoking Philomel, says:

> " And whiles against a thorn thou bear'st thy part
> To keep thy sharp woes waking."

Fletcher speaks of

> " The bird forlorn,
> That singeth with her breast against a thorn."

---

[1] See Ellis's "Specimens of the Early English Poets," vol. ii. p. 356.

And Pomfret, writing towards the close of the seventeenth century, says:

> "The first music of the grove we owe
> To mourning Philomel's harmonious woe;
> And while her grief in charming notes express'd,
> A thorny bramble pricks her tender breast."

The origin of such an odd notion it is not easy to ascertain, but I suspect Sir Thomas Browne was not far from the truth when he pointed to the fact that the Nightingale frequents thorny copses, and builds her nest amongst brambles on the ground. He inquires "whether it be any more than that she placeth some prickles on the outside of her nest, or roosteth in thorny, prickly places, where serpents may least approach her?"[1]

In an article upon this subject published in the "Zoologist" for 1862 (p. 8029), the Rev. A. C. Smith has narrated the discovery on two occasions of a strong thorn projecting upwards in the centre of the Nightingale's nest. It

---

[1] Sir Thos. Browne's Works, Wilkin's ed. vol. ii. p. 537.

cannot be doubted, however, that this was the result of accident rather than design; and Mr. Hewitson, in his "Eggs of British Birds," has adduced two similar instances in the case of the Hedge Sparrow.

The nest of the Nightingale is a very loosely-made structure, composed for the greater part of dead leaves, and placed upon a hedge bank, generally at the root of some stout shrub or thorn. The eggs, usually five in number, are, like the bird itself, of a plain olive-brown colour. The young Nightingales are spotted like young Robins, having the feathers of the upper portions of the plumage tipped with buff colour. In some respects the Nightingale assimilates very much in habits to the Robin; and advantage has been taken of this in localities where the Nightingale is unknown to introduce its eggs into the nests of Robins, with a view to having the young reared in the neighbourhood, and so induced to return to it. But although, as regards hatching and rearing, the plan has been successful, the birds have never returned to the place of their birth. For some inexplicable reason, a limit appears to be set to the migration of the Nightingale, which has no parallel in the case of other migrants.

As autumn approaches it moves southwards towards the Mediterranean, and spends the winter months in North Africa, Egypt, and Asia Minor. We cannot help thinking that the Nightingale and many other birds which visit us in summer and nest with us, must also nest in what we term their winter quarters; otherwise it would be impossible, considering the immense numbers which are captured on their first arrival, not only in England, but throughout central and southern Europe, to account for the apparently undiminished forces which reappear in the succeeding spring.

The late Mr. Blyth, however, was of a different opinion. Criticizing the above remarks, he wrote :—

"The only birds known to me that breed in their winter quarters are two species of Sandmartin (*Cotyle riparia* and *C. sinensis*). In India

I have been familiar enough with birds in their winter quarters, and have no hesitation in asserting that migratory species (with the remarkable exceptions named) do not even pair until they have returned to their summer haunts. Were they to do so, I could not but have repeatedly noticed the fact, and must needs have seen very many of their nests and young."

To my suggestion that from Mr. Layard's observation of young birds there, the Common Swallow, *H. rustica*, probably breeds at the Cape during the season that it is absent from the British Islands, Mr. Blyth replied:—

"According to my experience of *Hirundo rustica* (and I have had the best opportunities for observation), it decidedly does not breed in its winter quarters. Some birds of this species, which pass their non-breeding season within the tropics, may migrate south instead of north, and breed in the summer of the southern hemisphere instead of that of the northern hemisphere; but there is no reason to suppose that they are the same individuals. Were it so, the

Cape colony would indeed be flooded with *Hirundo rustica.* Besides, these birds renew their plumage (as the Cuckoo likewise does) when in their winter quarters; whereas the Sand-martins (*Cotyle*), as I am all but sure from recollection, resemble the great majority of our summer migrants in moulting before they take their departure equatorward. That our British Sand-martin (*C. riparia*) breeds in Egypt during the winter months is noticed in the ' Proceedings of the Zoological Society' for 1863 (p. 288), and that its ordinary representative in India and the countries eastward (*C. sinensis*) does the same I can vouch from personal observation, having myself taken both eggs and young about the turn of the year from their burrows in the banks of the Hugli; while Mr. Swinhoe noticed their breeding when in their winter haunts, in the ' Ibis ' for 1863, p. 257, and 1866, p. 134."

THE BLACKCAP.

(*Sylvia atricapilla.*)

FIVE species may be conveniently grouped under the generic term *Sylvia*, or Fruit-eating Warblers, and these, with one exception, visit Great Britain regularly in the spring. Two of them, the Blackcap and Garden Warbler, enjoy little more immunity from traps and birdlime than does the Nightingale. Their fine song marks them at once as the prey of the professional bird-catcher, and during the month of April immense numbers are taken daily. The Whitethroat and Lesser Whitethroat are also sought after as cage-birds, but not to the same extent, for their song is neither so musical nor so varied.

In no part of the country are these four species more plentiful than in the south-eastern counties of England; and the neighbourhood of the metropolis seems to have some special attraction for them. Far from shunning "the busy haunts of men," they appear to be nowhere more at home than in our gardens and orchards. The reason is obvious as soon as we become acquainted with their habits, and the nature of their food. We then discover that their motives are not so disinterested as we might suppose, since the real attraction is *fruit*. Upon this the parent birds live to a great extent; and after bringing up their young upon various kinds of insects which infest fruit trees—in which they unquestionably do us good service—they introduce their progeny at length to the more palatable pulp upon which they themselves have been faring so sumptuously. No wonder, then, that the large market-gardens of Kent, Surrey, and Middlesex should entice such numbers of these little birds to remain in their vicinity throughout the summer.

The Blackcap (*Sylvia atricapilla*) is the earliest of the genus to make his appearance, and seems to be hardier also than any of his congeners. Many instances are on record of Blackcaps having remained in this country throughout the winter, and this has been noticed as particularly the case in Ireland. It is rather singular that Mr. Yarrell, in referring to the sister isle, says that the Blackcap "has been taken, once at least, in the north of Ireland," as if he were of opinion that its occurrence there were doubtful, or at least extremely rare. Mr. Thompson, in his excellent " Natural History of Ireland" (vol. i. p. 183), notices the Blackcap as a regular summer visitant there; but he adds that it must be considered very local. In Scotland it is considered rare, being confined chiefly to the south; but since the observations were published from which these remarks are drawn, considerable changes seem to have taken place in the local distribution of many species of birds. This is notably the case with the Blackcap and Garden Warbler, both of which have followed cultivation, and now are found commonly in localities where twenty years ago they were either unknown or stated to be extremely rare.

The Blackcap, like the Nightingale, appears to migrate almost due north and south, and ranges from Lapland to the Cape. It is resident in Madeira, the Azores, and the Canaries, and is also found throughout the year in Northern Africa and Southern Italy. In the fine collection of African birds (*Passeres* and *Picariæ*) belonging to Mr. R. B. Sharpe, I have seen a specimen of the Blackcap from Senegal. In Spain and Portugal it is found only on the migration in spring and autumn. Mr. Godman, in his interesting work on the "Natural History of the Azores," has described a curious variety of the Blackcap which is found in these islands, "having the black marking on the head extending to the shoulders and round under the throat," and he was informed that individuals were sometimes found with "the whole of the under parts of the body black." This variety appears to have been met with also in Madeira, from whence it was described by Heineken ("Zool. Journ." v. p. 75). A figure of it will be found in Jardine and Selby's "Illustrations of Ornithology," pl. 94.

However much observers may be deceived by song, there is no mistaking either sex of the Blackcap as soon as the bird comes in view. The black crown of the male and the brown crown of the female suffice to distinguish the species amongst every other of our summer migrants. There is something very peculiar, too, about the half-hopping, half-creeping motions of all the Fruit-eating Warblers, which distinguishes them at once from other small birds frequenting the same haunts.

The males invariably arrive some days before the females; but both sexes seem to leave the country much about the same time—that is, early in September.

The nests of all the species in the genus *Sylvia*, as compared with those of the finches and linnets, are slovenly and loosely-made structures; and that of the Blackcap is no exception to the rule. The birds take some pains, however, to conceal it, and both male and female bestow a good deal of trouble upon it. It is generally placed a few feet from the ground, and is composed of dry bents, and lined with horsehair. The eggs, usually five in number, are white clouded with pale brown, and sparsely spotted with black towards the larger end. They closely resemble the eggs of the Garden Warbler, but differ in being smaller, and as a rule of a warmer tint; the pink or reddish-brown colour with which the eggs of the Blackcap are often suffused is not found in those of its congener. Both sexes take their turn at incubation, relieving one another to feed; but the male will often feed his partner on the nest, and then sit and sing to her. As to the song, it is simply delightful. I refrain, however, from attempting a description, for two reasons. The attempt has been made very often, and mere verbiage can convey but a very faint notion of its nature. It must be heard to be appreciated. If I were asked the question, "How am I to know the song when I hear it?" I would reply, "Approach the bird as slowly and as noiselessly as possible, until you can see the individual singing." This is the only way to learn the songs of birds. The note of each species then becomes impressed upon the memory, and can afterwards be detected without hesitation when the bird is not in sight. To acquire this knowledge, however, of the songs of birds, one thing is necessary—an ear for music. This, unfortunately, cannot be imparted by teaching; and unless it exist as a gift of nature, the delight of music can never be experienced. There is this consolation, however, for those who are not musicians—they cannot feel so much the loss of a pleasure which they have never experienced.

THE ORPHEAN WARBLER.

(*Sylvia orphea.*)

THE Orphean Warbler, as its name implies, is another noted song bird; but, though not uncommon in some parts of Europe and Asia, its claim to be included amongst our British warblers rests on very slender grounds. So long ago as July, 1848, a pair of this species were observed in a small plantation near Wetherby, and the hen bird was shot and forwarded to Sir William Milner, who informed Mr. Yarrell of the fact. On this single instance it was included by the last-named naturalist in his "History of British Birds." Since the last edition of that work was published (1856), there is reason to believe that the Orphean Warbler has occurred again at least on two occasions in England. In June, 1866, the late Sergeant-Major Hanley, of the 1st Life Guards, well known as a bird fancier, purchased a young warbler, which had been chased and caught by a boy near Holloway. Mr. Blyth, who saw it in the following December, pronounced it to be without doubt a female Orphean Warbler. As the bird when caught was unable to fly, it is evident that a pair must have nested in the neighbourhood. I have seen a nest and eggs which were taken in Notton Wood, near Wakefield, in June, 1864, which certainly appertained to none of our common warblers, and the eggs could not be distinguished from well-authenticated eggs of *Sylvia orphea*.[1] Mr. Howard Saunders has reported a similar nest and eggs from East Grinstead. The eggs differ from those of the Blackcap and Garden Warbler in being white, spotted, chiefly at the larger end, with ash-grey. The bird may be briefly described as a large form of the Blackcap, exceeding it by half an inch in total length, and by a quarter of an inch in length of wing, the male having the black crown which characterizes

---

[1] *Cf.* "Handbook of British Birds," p. 106.

our well-known songster, and resembling it generally in appearance. It differs, however, in having the bill shining black instead of horn colour, the under parts white instead of grey, the legs brown instead of slate colour, and the outer tail feathers margined with white instead of being uniformly grey. In habits and mode of life it assimilates, as might be expected, very much to the species with which we are so familiar. Those who have seen the nest, state that it is large for the size of the bird—a loose and open structure, rather shallow, and generally placed in a low bush near the ground. Mr. Yarrell has given very scanty information about this species, particularly as regards its geographical distribution, from which it might be inferred that very little is known of it. This, however, is not the case.

While the Blackcap migrates almost due north and south, the Orphean Warbler migrates westwards and northwards from the east and southeast, and *vice versâ*. In North-west India, particularly in the neighbourhood of Umballah, it is tolerably common. The Rev. Canon Tristram found it numerous in Palestine, and especially abundant under Mount Hermon. Messrs. Elwes and Buckley include it in their list of the birds of Turkey ("Ibis," 1870, p. 19), and Lord Lilford has noted its occasional occurrence in spring in the Ionian Isles. Rüppell includes it amongst the birds of Arabia and Egypt,[1] but either it is not very common in Egypt, or it has escaped the searching eyes of many English ornithologists in that country. Mr. O. Salvin found it tolerably common in the Eastern Atlas, and it has also been met with in Tripoli (*cf.* Chambers, "Ibis," 1867, p. 104). As it is thus found in North Africa, and, according to Professor Savi, is a summer visitant to Italy, one would naturally expect to find it in Malta; but Mr. C. A. Wright, who has paid great attention to Maltese ornithology for many years, states that he has never met with it himself, and that only one instance of its occurrence in Malta is

---

[1] "Syst. Uebers. d. Vögel Nord-Ost Afrika's," p. 57.

known to him. In Spain it has been observed as a summer visitant both by Lord Lilford and Mr. Howard Saunders. The last-named naturalist says ("Ibis," 1871, p. 212) that it nests there in May, and refers to the frequent inequality in the size of eggs in the same nest—a peculiarity which does not seem to have been previously noticed. In Portugal it appears to be only an occasional summer visitant, apparently not straying so far westward as a rule I am not aware that it has been found further to the south-west than Morocco. Mr. Tyrwhitt Drake met with it in this country in 1867, but considered it rare.

According to the observations of Von der Mühle, in his "Monograph of the European *Sylviidæ*," and of Captain Beavan on various birds in India ("Ibis," 1868, pp. 73, 74), there is good reason to believe that both the Blackcap and the Orphean Warbler completely lose the black crown in winter, and reassume it at the approach of the breeding season.

Criticizing these remarks, however, the late Mr. Blyth wrote:—

"Do the males of these birds lose the black cap in winter? Certainly not the former—at least as observed in captivity—and therefore I cannot help doubting exceedingly that they do so in the wild state. Upon a bad Indian drawing of the Orphean Warbler, reproduced in the 'Proceedings of the Zoological Society' for 1851 (p. 195, pl. 43), the supposed *Artamus cucullatus* was sought to be established. The habits of the Orphean Warbler are thus described in Jerdon's 'Birds of India'—in which country, by the way, it passes the winter, the males then retaining their black cap:—'It frequents groves, gardens, hedges, single trees, and even low bushes on the plains; is very active and restless, incessantly moving about from branch to branch, clinging to the twigs, and feeding on various insects, grubs, and caterpillars, and also on flower buds. It is sometimes seen alone, at other times two or three together.'

Undoubtedly it must needs feed also on soft fruits. The hen of this bird bears an exceedingly close resemblance to the Lesser Whitethroat, except in size; while the cock bird further differs in having the black cap *at all seasons*. There is likewise in India the *Sylvia*, or *Curruca, affinis*, which resembles our Lesser Whitethroat, excepting in being as large as our Common Whitethroat. The latter bird has lately turned up in the north-west of India; and the British Lesser Whitethroat is the only one of the group which extends its range eastward to Lower Bengal, where it occurs, however, only above the tideway of the rivers, upon the sandy soil in which the Baubul (*Vachelia farnesiana*) grows plentifully. There I have observed our familiar little friend in abundance during the winter months, but never upon the alluvion or mud soil; and the same remark applies to *Hippolais rama*. It has been suggested to me that there may be a race of 'Blackcap' that visits Eastern Europe, the males of which have a rufous-brown cap like the females. In our race of Blackcap the diversity of the sexes is very noticeable, even in nestlings."

Captain Beavan, in the article before referred to, says: "Specimens of the Orphean Warbler, procured on the 22nd of October, had no trace whatever of a black head, and were considered by Colonel Tytler to be the young of the year; but in my opinion the state of the plumage was not sufficiently juvenile; and I think that the old birds adopt a different colouring according to the time of year, probably putting on the black head as the breeding season approaches." To this observation the editor of the "Ibis" appended the following note: "That this view of the case is correct there is probably little doubt (*cf.* Von der Mühle, 'Monogr. Europ. Sylv.,' p. 48)."

From these observations it was surmised that the same might be the case with the Blackcap.

THE GARDEN WARBLER.

(Sylvia hortensis.)

TO those who are unacquainted with the bird, the Garden Warbler may be best described as equal in size to the female Blackcap, resembling it in colour without the chestnut crown, having the belly pure white instead of greyish white, and the legs lighter in colour. It appears much later than the Blackcap, seldom arriving before the end of April. Both sexes are alike in outward appearance; but it has been ascertained, by careful observers who have dissected the birds, that the males invariably arrive in this country before the females. Pennant, Montagu, and other old authors, called this bird the Greater Pettychaps, while they bestowed the name of Lesser Pettychaps—presumably from its resemblance in miniature—upon the Chiffchaff.

Throughout England the Garden Warbler appears to be pretty generally distributed. Mr. A. G. More, however, in his essay on the Distribution of Birds in Great Britain during the nesting season ("Ibis," 1865, p. 25), speaks of it as scarce in Cornwall and Pembrokeshire, and absent from Wales. Mr. Rodd, on the other hand, characterizes the Garden Warbler as a summer visitant to East Cornwall, and says it "breeds annually in the woods at Trebartha, in North Hill, from whence specimens of its nest and eggs have been received."[1] He adds also that it has once been met with near Penzance; and that in the autumn of 1849 several speci-

---

[1] See "List of British Birds, as a Guide to the Ornithology of Cornwall," 2nd edition, 1869, p. 15.

mens were obtained from Scilly. Dr. Bullmore, in his "Cornish Fauna" (p. 17), confirms Mr. Rodd's statement that it is a summer visitant to East Cornwall.

It will be remarkable if this bird is not found to be common in some parts of Wales, since it not only occurs in Ireland, but is not nearly so scarce there as the observations of Mr. Thompson would lead us to suppose. In his "Natural History of Ireland" (Birds, vol. i. p. 185), this naturalist refers to the Garden Warbler as extremely rare in Ireland, and notices its occurrence only in the counties of Cork and Tipperary. If I mistake not, Mr. Blake-Knox has met with it in the county of Dublin; I have myself observed it in Wicklow; and Sir Victor Brooke has lately assured me that in the county of Fermanagh, about Lough Erne, it is common in summer, and nests regularly in the neighbourhood of Castle Caldwell, to the northwest of that county. In the same neighbourhood, he added, the Blackcap is unknown. When we remember the number of naturalists with whom Mr. Thompson was in correspondence in all parts of Ireland, it is singular that so few of them should have been able to report the presence of this bird in their respective districts. I have already referred to the changes which have taken place in the local distribution of many species of birds within the last twenty or thirty years, and there is no reason for doubting that the statements published by Mr. Thompson in 1849, and the observations of naturalists of the present day, are both perfectly correct, and that the Garden Warbler, like many other birds, is now common in localities where formerly it was unknown. The number of resident naturalists in Wales is very small as compared with England; nevertheless, it is to be hoped that those who have the opportunity will examine into the truth of the alleged absence from Wales of this bird, and publish the result of their investigations.

The limit of the Garden Warbler's range northwards in the British Islands has not been satisfactorily ascertained. That it is found in many parts of the south of Scotland we know from the observations of Macgillivray and the late Sir William Jardine; but we have yet to learn whether it penetrates to the Highlands or visits the Hebrides. According to Selby, it is found throughout the greater part of Scotland; but Mr. Robert Gray, in his recently published " Birds of the West of Scotland," is disposed to think that it is not commonly distributed. It is, as he says, very difficult to judge of the comparative numbers of so shy a bird, as it is even less frequently noticed, save by the patient observer, than some other species of greater rarity. " In the sheltered and wooded districts of the midland and southern counties," he adds, " it is one of the most attractive songsters, tuning its loud and gleeful pipe on the top of some fruit tree an hour or two after daybreak, and again about the dusk of the evening. These love notes, however, are not of long continuance, for the bird becomes silent after the young are hatched, unless a second brood is reared, when the same wild yet mellow blackbird-like song is again for a short time heard. Mr. Sinclair has observed the Garden Warbler at Inverkip in Renfrewshire, where the richly-wooded preserves afford it a constant shelter during its summer sojourn." In Shetland, according to Dr. Saxby,[1] it is a rare autumn visitor, usually occurring in September. By exercising great caution he has sometimes approached within a few feet of the bird, and watched it picking the green *aphides* from the sycamore leaves. It does not appear to have been observed in Orkney. Its range northwards in Europe, according to Nilsson, extends to Sweden, where it is observed to be a regular summer visitant, arriving in May and leaving in August. In all the countries bordering the Mediterranean it appears to be well known. Mr. Saunders informs me that it is common in Spain in spring and autumn; and Mr. Wright, referring to its presence at the same seasons in Malta, where it is

---

[1] "The Birds of Shetland," p. 73.

known as the far-famed "beccafico" of the Italians, says that as many as a hundred dozen are sometimes brought in at a time.[1] Lord Lilford has once found this bird nesting in Epirus.[2] The late Mr. C. J. Andersson met with it as far south as Damaraland, South-west Africa. In habits the Garden Warbler closely resembles other members of the genus. Shy and restless, it differs from the Blackcap in its inferior powers of song, and from the Whitethroats in being less garrulous. It is nevertheless a beautiful songster, and will sometimes sit in the midst of a thick bush in the evening, like a Nightingale, and maintain a continued warble for several minutes without a pause. Its song is somewhat irregular, both in time and tune, but it is wonderfully mellow for so small a bird. It sometimes commences its song like a Blackbird, but always ends with its own. In some of its actions it resembles the Willow Wren, for it seems constantly in motion, hopping from bough

---

[1] "Ibis," 1864, p. 67.   [2] "Ibis," 1860, p. 231.

to bough in search of insects, and singing at intervals. It is very partial to fruit of all kinds, but at the same time destroys vast numbers of caterpillars, spiders, and *aphides*. Much against my inclination I have shot a few Garden Warblers in the spring soon after their arrival, for the purpose of ascertaining the nature of their food, and can therefore affirm, from personal inspection, that they destroy quantities of insects which are destructive to foliage. Under the head of Blackcap, I have referred to the nest of the Garden Warbler for the purpose of comparison, and need only add here that it is generally well concealed, and that, unless the owner is seen near the nest, it is oftentimes not very easy to distinguish the eggs from those of its congener, which have been already described.

THE COMMON WHITETHROAT.

(*Sylvia cinerea.*)

FAR from leading a retired life like the last-named bird, the Whitethroat forces itself into notice by its noisy chattering and repeated sallies into the air. We cannot walk along a country lane in May without being reminded at every twenty yards of the presence of this demonstrative little bird. With crest-feathers erect and half-extended wings, it bustles in and out, gesticulating loudly, and seems to live in a perpetual state of excitement.

The country lads call it the "Nettle Creeper," from its frequenting overgrown ditches and hedgebanks where the nettle is plentiful, amongst the stems of which it builds its nest. It comes to us about the third week in April, and remains until the end of August. It is very generally distributed in the British Islands, and is as common in Ireland as it is in England. In the north of Scotland it is said to be rare; but a correspondent of Mr. More finds it breeding regularly in Mull and Iona.[1] It visits Scandinavia in summer, and is found also at that season in Russia and Siberia. It is one of the commonest birds in spring and autumn in Malta, and is occasionally observed in Corfu in September and October. In winter it is not uncommon in Asia Minor and North-east Africa. Amongst the birds collected at Aboo, North-west India, by Dr. King, in September, 1868, Mr. Hume found one which both he and M. Jules Verreaux identified at

---

[1] *Cf.* "Ibis," 1865, p. 25.

once as *Sylvia cinerea*. Unlike the Garden Warbler, the Whitethroat sings a good deal on the wing, sometimes returning to the branch it has just left, after the manner of a Tree Pipit, sometimes re-alighting elsewhere. The song, which is commenced on arrival, generally ceases early in the month of July. Its habits, and as Mr. Thompson says, the grotesquely earnest appearance which the erected crest, feathers, and distended throat impart when singing, render this bird one of the most interesting of our warblers. It seems to prefer the tallest and thickest hedgerows, where there are plenty of brambles and briars, and ditches which are choked with weeds and nettles. It does not keep, however, to the fields and lanes, but visits our gardens and orchards in company with its young to pilfer currants, raspberries, and other fruit when ripe. The caterpillars to be found on the currant trees are favourite morsels with this bird, and we should not forget that if it takes a few currants it is also the means of saving a good many.

The nest of the Whitethroat is generally placed near the ground, amongst nettles or other rank herbage, and is constructed of dry grass-stems and horsehair. The eggs, usually five in number, are minutely speckled all over with ash-brown or ash-green, and spotted at the larger end with gray. I have watched an old Whitethroat bringing food to its young, and have been surprised to see in how short a space of time it contrived to find food and return to the nest. Sometimes it was impossible to see even with a glass what this food was, but at other times I could plainly discern a caterpillar wriggling between the mandibles.

THE LESSER WHITETHROAT.

(Sylvia sylviella.)

THIS is not nearly so common a bird, nor so generally distributed in Great Britain, as the last-named. It is confined more or less to the midland and southern counties of England, is very rare in Scotland, and unknown in Ireland. Mr. Rodd, in his "List of Birds" before quoted, says the Lesser Whitethroat is only seen in Cornwall during the autumn migration, and then only occasionally at Scilly. In Wales it appears to be equally scarce (*cf.* More, "Ibis," 1865,

p. 25), but it is possible that, from its general resemblance to the last-named bird, it may have been often overlooked. The respective measurements of the two species are as follows :—

	Total length.	Wing.	Tarsus.
Common Whitethroat	5·5 in.	2·9 in.	·8 in.
Lesser Whitethroat	5·2 ,,	2·5 ,,	·7 ,,

Independently, however, of its smaller size, the Lesser Whitethroat may be distinguished by its black ear-coverts, and by the absence of the pale rufous edgings to the secondaries, which are so conspicuous in the larger species. The legs also are slate-coloured instead of yellowish-brown.

In haunts, habits, and mode of nesting the two species are very similar, and what has been said of one will apply almost equally well to the other. Both arrive also about the same time—namely, the third week in April; and by the end of August, when the young are strong enough to shift for themselves, they depart again southwards. Although the nests of the two species are very similar, the eggs of the Lesser

Whitethroat have a much clearer ground-colour, and are never so profusely freckled as those of its congener. On the contrary, the spots of ash-brown, or ash-green, are almost always at the larger end, leaving the smaller end of the egg almost spotless.

The range of the Lesser Whitethroat southward is probably more or less identical with that of the Common Whitethroat. It is abundant in Spain in winter and early spring, but does not remain to breed there. In Malta, strange to say, it has only been recognised once; but in Egypt and Nubia, especially from Dendera to the First Cataract, it is very numerous in winter. Individuals of this species have been seen to alight on vessels in the Mediterranean, even when upwards of sixty miles from the nearest land, and thus its ability to migrate from Europe to Africa, and back, is sufficiently established. Eastward it penetrates to Lower Bengal, where, in the cold season, it is said to be not uncommon.

THE REDSTART.

(Ruticilla phœnicura.)

SPRIGHTLY in its actions, and more vividly coloured than many of our Summer Migrants, the Redstart cannot fail to attract attention in the districts which it frequents during its sojourn with us. It would be difficult, indeed, to find a more beautiful little bird than the male Redstart in his nuptial plumage. The pale grey colour of the head and back, relieved by a silvery white spot upon the forehead and a jet-black throat, contrasts strongly with the bright chestnut of the breast, upper tail coverts, and tail. From the bright colour of its tail, in fact, it has derived the name Redstart, which is simply the Anglo-Saxon equivalent for " Redtail." " Fire-tail," " Brand-tail," and " Quickstart," are other local names by which it is variously known. The last-named has reference to the singularly characteristic movement of the tail, which is rapidly flirted horizontally instead of vertically, as in the case of most other birds.

Upon this point, however, there seems to be some difference of opinion. Macgillivray, a high authority in such matters, observes, " As to the motion of the tail in this bird, which has supplied some observers with a subject of dispute, I am convinced that it is vertical—that is, up and down, and not alternately to either side, although at each jerk the feathers are a little spread out, as is the case with those of many other birds of this order, as the Stonechat and Whinchat." I feel sure, notwithstanding this opinion, that I have frequently observed a horizontal movement.

Its mode of progression on the ground has been compared by the same observer to that of the Wheatear, "for it neither walks nor runs," he says, "but advances by leaps." I cannot, however, completely endorse this view, for I have frequently seen a Wheatear run, and at times very rapidly. " Unless on a wall, or on bare ground, however, it seldom hops much, for it procures its food chiefly by sallying after insects on the wing, or by alighting on the ground to pick up those which it has observed amongst the herbage, and on trees it flies from branch to branch."

Although generally distributed in England and Scotland, the Redstart is nowhere very common, being most plentiful, apparently, in the southern counties of England, and becoming rarer as we proceed northward. In Ireland it is scarcely known at all, and does not visit the Hebrides. On the Continent, however, it has a tolerably wide range, extending from Archangel throughout Scandinavia and the whole of Europe, except Portugal, to the Mediterranean, which it crosses to visit North Africa, Egypt, and Abyssinia for the winter season.

The haunts which it affects in this country are generally not far removed from human habitation, and it is not unusual to find the nest, containing five or six pale-blue eggs, upon a peach or plum-tree against a wall; upon a cross-beam of a summer-house; or in a hole of a wall or tree, as opportunity may serve. The eggs are very similar to those of the Hedge Sparrow, but are invariably smaller and paler. It picks up most of its food, such as small beetles, spiders, and worms, on the ground; and its actions when thus engaged remind one more of the Robin than of the Wheatear, as Macgillivray thought. At other times it will sit upon an exposed branch, and dart forth into the air, like a Flycatcher, to secure a passing insect. Its song, though sprightly, is weak and seldom prolonged. It is generally poured forth from some bough or other "coign of vantage," but is occasionally uttered as the bird hovers on the wing, or flies from spray to spray.

Although a very shy bird, the Redstart occasionally takes up its quarters close to the house, and when once it has selected a site for its nest and hatched its young, it manifests such attachment for them as to allow a very near approach, and will even permit a visitor to stroke it as it sits upon the nest.

The beauty of its plumage, its sprightly and at times incessant song, and the good which it effects in ridding plants and fruit-trees of the green *aphis*, commend it to the notice and protection of all owners of gardens.

The Common Redstart has scarcely quitted our shores in autumn before its congener, the Black Redstart (*Ruticilla tithys*), arrives to pass the winter here, and occasionally even to linger on until the more familiar species returns again with the spring. But since it is properly regarded as a winter visitant to this country, any lengthened description of the species, and of its haunts and habits, would be out of place here. I shall therefore merely observe that it may be distinguished from the Common Redstart by the sooty-black colour of the breast and belly, which parts in the other are orange-brown, and that it generally arrives about the first week in November, and remains until the end of March or beginning of April.

The origin of the specific name "*tithys*" seems to be somewhat doubtful, although several ornithologists have attempted an explanation. Hemprich and Ehrenberg ("Symbolæ Physicæ," fol. bb), and Von Heuglin ("Orn. Nord-Ost Afrika's," i. p. 334) have referred it to τίτης, *ultor*, with which, however, in the opinion of Professor Newton ("Ann. Mag. Nat. History," Ser. 4, x. p. 227), it can have nothing to do. Professor Newton himself, in the magazine just quoted, and in a footnote to his edition of Yarrell's "History of British Birds," i. p. 333, writes: "*Sylvia tithys* (by mistake) Scopoli, Annus I. Historico-naturalis, p. 157 (1769). This naturalist admittedly took his specific name from Linnæus, who spelt the word '*titys*' as did Gesner; but the best classical authorities, Stephanus, Porson, and Passow, consider '*titis*' to be right. This originally meant a small chirping bird, and is possibly cognate with the first syllable of our *tit*mouse and *tit*lark." After the opinion expressed by such authorities, it may appear somewhat presumptuous on my part to offer a suggestion; but there is yet another explanation, which has apparently been overlooked. Might not the word "*tithys*" (more correctly "*tithus*") be derived from the Greek adjective τιθός, θή, θόν, which has the same signification as τιθασός, that is, "reared up in the house, domesticated." Compare the domestic hens of Dioscorides, τιθαὶ ὄρνιθες. The term "domesticated" would be well applied to the Black Redstart, which is a very familiar bird, frequently perching on house-tops and garden walls, and building in holes and crannies in the neighbourhood of man's dwelling.

THE SEDGE WARBLER.

(Salicaria phragmitis.)

LEAVING the woods, gardens, and plantations, and proceeding to the river side, we meet with a very different class of birds—the river warblers. This is a very numerous family, and were we about to treat of all the known species, it might be advisable for simplicity's sake to group them into sub-families. As we are confining our attention, however, for the present, to those species only which have been met with in the British Islands, it will be less confusing if we dispense with this subdivision, and notice them under the same generic name—*Salicaria*. The various members of this genus may be distinguished by their short wings, rounded tails, tarsus longer than the middle toe, large feet, long and curved claws, and large hind toe with strong curved claw. They differ, too, from other warblers in their habit of singing at night. There are eight species which have all more or less a claim to be included in the British list, although three only can be regarded as regular summer migrants. These three are the Sedge Warbler (*S. phragmitis*), the Reed Warbler (*S. strepera*), and the Grasshopper Warbler (*S. locustella*). The others are Savi's Warbler (*S. luscinoides*), the Aquatic Warbler (*S. aquatica*), the Marsh Warbler (*S. palustris*), the Great Reed Warbler (*S. arundinacea*), and the Rufous Warbler (*S. galactoides*).

The Sedge Warbler and the Reed Warbler generally arrive much about the same time in April, but, from some unexplained cause, the latter is much more restricted in its distribution than the former. The Sedge Warbler is found throughout the British Islands, but the Reed Warbler is almost unknown in Ireland, and its nest has only once been met with in Scotland.[1] As a rule, it is seldom, if ever, to be seen further north than Yorkshire and Lancashire, and does not breed either in Devon or Cornwall. It may thus be said to be almost confined to the eastern, midland, and south-eastern counties of England. Beyond the British Islands, too, it is less erratic in its movements than its congeners. The Sedge Warbler visits Scandinavia, Russia, and Siberia, and is found throughout Europe in summer, and in North Africa and Asia Minor in winter. The late Mr. Andersson sent specimens even from Damaraland, S.W. Africa. The Reed Warbler does not migrate as far north as this; but Mr. Gurney has received a specimen from Natal; and if we may rely on the identification of specimens obtained by Mr. Hodgson, it ranges as far eastward as Nepal.

---

[1] This was in Haddingtonshire, by Mr. Hepburn. See "Ibis," 1865, p. 24.

I have sometimes heard persons express their inability to distinguish these two species apart; but there ought to be no difficulty in the matter. The Sedge Warbler has a variegated back, with a conspicuous light streak over the eye; the Reed Warbler has a uniform pale-brown back, and the superciliary streak very faint. The actions of the two birds are not unlike, but their nesting habits are very different. *S. phragmitis* builds on the ground or very near it, making a nest of moss and grass, lined with horsehair, and laying five or six eggs of a yellowish-brown colour, with a few scattered spots or lines of a darker colour at the larger end. *S. strepera* suspends its nest between reed stems or twigs, round which a great portion of the nest is woven, and the entire structure is much larger, deeper, and more cup-shaped. The materials are long grasses, flowering reed-heads, and wool, the lining being composed of fine grass and hair. The eggs, five or six in number, are greenish-white speckled with ash-green and pale-brown. The habit which the Reed Warbler has of occasionally nesting at a distance from water is now probably well known to ornithologists. It was noticed by Mr. R. Mitford in the "Zoologist" for 1864 (p. 9109), and subsequently by the writer, in "The Birds of Middlesex," 1866 (p. 47), and by the author of "The Birds of Berks and Bucks," 1868 (p. 81). Mr. B. Hamilton Booth, of Malton, Yorkshire, communicated the fact of his having discovered a nest of the Reed Warbler in a yew tree, built so as to include three or four twigs as if they were reeds, and placed at a height of at least twelve or fourteen feet from the ground. He accounted for the nest being built at such a height, and in a tree, on the supposition that the first nest had been destroyed by the rats which infest the place, and the birds had taken a precaution for future safety.

THE GRASSHOPPER WARBLER.

(Salicaria locustella.)

THE third species of this genus which is a regular summer migrant to this country is the Grasshopper Warbler, so called from its peculiar sibilant note. In its general appearance it is most like the Sedge Warbler, but is larger in every way, and has the upper part of the plumage more variegated, no superciliary streak, and the throat minutely spotted. This last feature, however, is peculiar to the male. In habits, haunts, and in the character of its nest and eggs, the Grasshopper Warbler differs entirely from the two species above mentioned. It delights in a dense undergrowth or thick hedge-bottom, where it creeps about more like a mouse than a bird, and is extremely difficult to catch sight of, pausing at intervals to seize an insect or to give forth its remarkable note. Its well-made and compact nest, so different from the slovenly structure of the Sedge Warbler, is placed upon the ground, and carefully concealed. The eggs, five or six in number, are amongst the most beautiful of small birds' eggs. When blown they are white, minutely freckled over with brownish-red; but before the yolk has been expelled they are suffused with a delicate rosy tint, which afterwards unfortunately disappears. The Grasshopper Warbler is a regular summer visitant to Ireland, and is also found in the south of Scotland. Its retiring habits probably cause it to be overlooked, and were it not for its loud note it would doubtless often escape notice altogether. It does not appear to be anywhere a numerous species, and its geographical distribution has not been yet clearly defined. It is observed in Southern Europe at the periods of migration, and we may therefore presume that it accompanies its congeners and other small summer migrants to North Africa, Asia Minor, and Palestine.

SAVI'S WARBLER.

(*Salicaria luscinoides.*)

BEFORE the fens were drained, it is said that the rarer species, Savi's Warbler, was not uncommon in the eastern counties of England. The fen-men used to distinguish it from the Grasshopper Warbler by its note, calling the commoner species "the reeler," the other "the night reeler," from the resemblance of its note to the whirr of the reel used by the wool-spinners. In Norfolk, according to Mr. Stevenson, it appears to have been known to the marsh-men as "the red craking reed-wren." The fens of Baitsbight, Burwell, and Whittlesea were formerly noted localities for this species, then regarded as a regular summer migrant; but extensive drainage and increased cultivation of waste land has apparently destroyed the only breeding haunt which had any attraction for it, and it can now be only considered a rare summer visitant. I have once, and only once, seen this species alive in England. This was in a large reed-bed close to the river, near Iken, in Suffolk, in the month of September, 1874. The bird first attracted my attention by the very rufous colour of the dorsal plumage, and as I succeeded in obtaining a near view of it, I feel confident that I was not mistaken in the species. The nest and eggs of this bird are reported to have been taken in Norfolk, Cambridge, Huntingdon, Essex, Kent, and once in Devonshire.[1] In general appearance at a distance it is not unlike the Reed Warbler, but on closer inspection will be found to have the upper portions of the plumage and the tail more rufous, like the

---

[1] "Ibis," 1865, p. 23.

Nightingale; hence the term *luscinoides* which has been applied to it. The English name is borrowed from its discoverer, Signor Savi, who found it in Tuscany, and published an account of it in the "Nuovo Giornale di Litteratura," 1824, and in his "Ornithologia Toscana," vol. i. p. 270. The eggs are something like those of the Grasshopper Warbler, but larger and darker; the nest is very different, being composed entirely of sedge, so closely woven and interlaced as to remind one of the mat-baskets which are used by fishmongers. Of the geographical distribution of this bird we have yet a good deal to learn. It does not appear to range very far northwards, but is observed annually in summer in Southern Europe, passing by way of Sicily and the Maltese Islands to Egypt. Mr. Salvin found it abundant in the Marsh of Zana, and Mr. Tyrwhitt Drake met with it in Tangier and Eastern Morocco.

THE AQUATIC WARBLER.

(*Salicaria aquatica.*)

ON three occasions only has the Aquatic Warbler been recognised in England. One taken at Hove, near Brighton, in October, 1853, is in the collection of Mr. Borrer;[1] a second, in my possession, was killed near Loughborough, in the summer of 1864;[2] and a third, believed to have been obtained near Dover, is in the Dover Museum.[3] This bird resembles the Sedge Warbler in size and general appearance, but, in addition to the light stripe over each eye, it differs in having a light stripe down the centre of the forehead; this, being very distinct, furnishes a good means of identifying it readily. The species has been figured by Dr. Bree in his "Birds of Europe," to which work the reader may be referred for

---

[1] *Cf.* Newton, P. Z. S., 1866, p. 210.
[2] "Ibis," 1867, p. 468.
[3] *Cf.* J. H. Gurney, jun., "Zoologist," 1871, p. 2521.

further information and a more detailed description. I may supplement his remarks, however, by saying that Lord Lilford found it common in Corfu in May, and at Nice in August and September;[1] and that Mr. T. Drake met with it in March in Tangier and Eastern Morocco.[2] Now that its occasional presence in this country has been detected, ornithologists should look out for it between April and September, and scrutinize every Sedge-bird they see, on the chance of meeting with the rarer species.

THE MARSH WARBLER.

(Salicaria palustris.)

IN appearance this bird resembles the common Reed Warbler, just as the Aquatic Warbler resembles the Sedge-bird. It is one of the plain-backed species, and similarity in appearance as well as in habits causes it doubt-

---

[1] "Ibis," 1860, p. 232.   [2] "Ibis," 1867, p. 426.

less to be overlooked or mistaken for the commoner bird.

From its general resemblance to the Reed Warbler, *Salicaria strepera*[1] (Vieillot), it has no doubt been overlooked; but when its distinguishing characters have been duly noted it will in all probability be found to be a regular summer migrant to this country. Dr. Bree, when treating of this species in his "Birds of Europe," says (vol. ii. p. 74): "I think it very probable that this bird is an inhabitant of Great Britain, though hitherto confounded with the Reed Warbler. I think I have myself taken the nest; and Mr. Sweet's bird, mentioned by Mr. Yarrell, was probably this species."

In the "Zoologist" for 1861, p. 7755, the occurrence of the Marsh Warbler in Great Britain was recorded by Mr. Saville, who procured a single specimen, subsequently identified by Mr. Gould, and saw others in Wicken Fen,

---

[1] The specific name *arundinacea*, which is commonly applied to this species, belongs properly to the Great Reed Warbler, the *Turdus arundinaceus* of Linnæus.

Cambridgeshire. He says: "My attention was first attracted to this species some time since, during a visit to our fens, by the marked difference in the song of a bird somewhat similar in appearance to the true *S. arundinacea* (i. e., *strepera*); it was louder, clearer, and sweeter-toned than that of the last-named. Its mode of flight, too, was more undulated and quicker. It was more shy and timid, continually retreating to the thickest covert. Never, so far as my experience goes, does it emit notes similar to the syllables 'chee-chee-chee' so common to *S. arundinacea*."

Another specimen of this bird was obtained in Cambridgeshire by the late James Hamilton, jun., of Minard, during the summer of 1864, and was exhibited at a meeting of the Natural History Society of Glasgow in February, 1865, as recorded by Mr. E. R. Alston in the "Zoologist," 1866, p. 496.

In the same year, Mr. Robert Mitford gave an account ("Zoologist," 1864, p. 9109) of a Reed Warbler which he found nesting in lilac trees in his garden at Hampstead, and which at the time was thought to differ specifically from *S. strepera*, and possibly to be *S. palustris*. In the summer of 1863 Mr. Mitford had found four pairs of this bird breeding in gardens under similar circumstances, and in July, 1865, he shot two of the same birds, both males, and found, as he says, " two nests similar in structure, and similarly situated to those of the previous year in my garden, from both of which the young had evidently flown only a few days previously. The birds were not in good order, but just beginning their moult. I so arranged the matter that at the time I shot these birds I received from Romney Marsh fresh-killed specimens of the true Reed Warbler, shot in the reeds of the fen ditches; and in comparing the two birds in the flesh together, I have little hesitation in saying that the inland warbler is not our Reed Warbler. I will not enter into the chief points of difference at present, as I hope next May to get a specimen or two in finer plumage." (" Zoologist," 1865, p. 9847.)

Mr. Mitford I believe has not altered the opinion which he originally expressed; but, from a careful examination of the birds shot by him, I am inclined to regard them all as *S. strepera*. This peculiarity in the Reed Warbler of nesting at a distance from water has since been noticed by naturalists in other parts of the country. In 1866 I referred to a confirmation of the fact as communicated by a friend at Ealing,[1] and Mr. A. C. Kennedy, in his " Birds of Berks and Bucks" (p. 81), has alluded to the same habit from his own observation near Windsor. In all probability the birds seen by Lord Clermont in lilac bushes at Twickenham[2] were also Reed Warblers.

Mr. Frederick Bond some time since called my attention to the occurrence of the rarer *S. palustris* in Norfolk, and kindly lent me a series of skins of both species procured in Cambridgeshire, Norfolk, and Sussex. Of these, two specimens of *S. palustris* were killed at Whittlesford, Cam-

---

[1] " The Birds of Middlesex," p. 47.
[2] " Zoologist," 1865, p. 9729.

bridgeshire, many years ago, under the impression that they were *S. strepera;* and three others near Norwich in June, 1869, under the like misapprehension. They do not differ in any way from skins of *palustris* from France and Germany, with which I have compared them.

The characters by which this species may be distinguished from *S. strepera* may be briefly stated as follows :—

Although the colour of the upper portion of the plumage in both is a uniform olive-brown, *S. palustris* is yellower. It is a somewhat longer bird, with a shorter and broader bill; a buffy-white line, extending from the base of the bill over the eye, is clearly defined. In *strepera* this line is so faint as to be scarcely discernible. Mr. Yarrell, indeed, considered it to be absent in *strepera;* but, from this circumstance, and from the fact of his describing the legs of this species as pale-brown, it may be inferred that he had before him, and figured, a young bird.

The first primary in the wing of both is very

short, quite rudimentary, in fact; while the third in each is the longest in the wing. In *palustris* the second primary is equal to the fourth; while in *strepera* the second is equal to the fifth. It is doubtful whether this can be invariably relied upon, for the length of feathers, even in the same species, will sometimes vary considerably, through age, moult, or accident.

The readiest means of distinguishing the two birds at a glance will be by the colour of the legs and toes. In living or freshly-killed specimens it will be observed that the tarsi and feet of *strepera* are of a slaty-brown colour, while in *palustris* the same parts are flesh-colour. In dried skins, the former turns to hair-brown; the latter to yellowish-brown. The tarsus of *palustris*, moreover, is rather longer and stouter than that of its congener. From this it appears that Mr. Gould in his " Birds of Great Britain " has figured *palustris* for *strepera*.

Dr. Bree, in his " Birds of Europe," has unfortunately figured *palustris* with slate-coloured legs and feet, which quite alters its appearance, although he has been careful in the text to describe the colour correctly.

The tail in *palustris* is less rounded than in *strepera;* the outer tail-feather in the former being not so short as in the latter.

The measurements of the two species, taken from skins, are as follows :—

	Length.	Bill.	Wing from carpus.	Tarsus.
S. strepera	5·3 in.	0·5½	2·7	0·8
S. palustris	5·5 in.	0·5	2·5	0·9

The nests and eggs differ as much as do the birds themselves.

The nest of *palustris* is much neater and more compact, and, as regards depth, not more than half the size of that of *strepera*. The eggs of both are subject to variation ; but, as a rule, it may be said that in those of *palustris* the white ground colour has little if any of the greenish or brownish tinge with which those of *strepera* are invariably suffused.

I have seen two nests in the collection of Mr. Bond, one containing three, and the other two eggs, taken at Whittlesford, which I have no doubt belonged to *palustris*.

In Badeker's work on the eggs of European birds, it is stated that the Marsh Warbler "builds in bushes, in meadows, and on the banks of ditches, rivers, ponds, and lakes. The nest is made of dry grass and straws, with panicles, and interwoven with strips of inner bark and horsehair outside. The rim is only very slightly drawn in. It has a loose substructure, and is by this and its half globular form, suspended on dry ground between the branches of the bushes or nettles, easily distinguished from the strongly formed nest of *S. strepera*, which is moreover built over water.[1] It lays five or six eggs the beginning of June, which have a bluish-white ground, with pale-violet and clear brown spots in the texture of the shell, and delicate dark brown spots on the surface, mingled with which are a number of black dots. The ground colour also in many fresh eggs is green, but

---

[1] Not always, as shown above.

clear, and very different from the muddy tint of the egg of the Reed Warbler. The female sits daily for some hours; but the male takes his turn. Incubation lasts thirteen days."

It would be extremely satisfactory to establish the fact of the regular migration to this country in spring of the Marsh Warbler; and it is to be hoped that ornithologists in all parts of the kingdom will not omit to investigate the subject, and record their observations.

THE GREAT REED WARBLER.

(Salicaria arundinacea.)

NOT only has this fine species visited England on several occasions, but in a few instances it has been found nesting here. It has, therefore, a good claim to be introduced into the present sketch. Specimens of the bird have been obtained, once in Northumberland, and three or four times in Kent,[1] and the eggs have been taken in Hertfordshire and Northamptonshire.[2] The reader has only to picture to himself a bird like the Reed Wren, but twice its size, and he will have an idea of the appearance of the Great Reed Warbler. Nor does the resemblance end here. It makes a nest just like the Reed Wren, but much larger, and lays eggs similarly coloured, but larger. It is a fine species, and its loud and varied notes, when once heard, can never be forgotten. Those who have had opportunities, such as I have enjoyed, on the opposite shores of Holland, of listening to this bird will regret with me that its visits to England are not more frequent. It is possible, as suggested by Mr. Hancock in the earliest notice of its occurrence here,[3] that it may be a regular summer visitant to our island; but its song is so loud and so remarkable, that I cannot think it could escape the notice of

---

[1] *Cf.* Yarrell, "Hist. Brit. Birds," vol. i. pp. 300, 301.
[2] *Cf.* "Ibis," 1865, p. 24.
[3] "Ann. Mag. Nat. Hist." 1847, p. 135.

any naturalist. The species is tolerably well dispersed throughout Europe, and according to Mr. Yarrell has been found as far eastward as Bengal, Japan, and Borneo. The Eastern bird, however, would appear to be the *Salicaria turdoides orientalis* of the "Fauna Japonica," and distinct from the European species. See Captain Blakiston on the Ornithology of Northern Japan, " Ibis," 1862, p. 317 ; Mr. Swinhoe on Formosan Ornithology, " Ibis," 1863, p. 305 ; and the Rev. H. B. Tristram, " Ibis," 1867, p. 78, on the Ornithology of Palestine, where both forms occur.

THE RUFOUS WARBLER.

(*Aedon galactodes.*)

FROM its peculiar coloration this bird is not likely to be confounded with any other species. Apart from the rufous tint of the upper portion of the plumage which has suggested its English name, the tail is totally unlike that of any of the river warblers; for, instead of being of a uniform brown, it has a broad band of black across both webs of all the feathers (except the two centre ones) towards their extremities, which black band is terminated by white. This is very conspicuous as the bird moves it up and down, and could not fail to attract the notice of anyone who has paid attention to birds. It does not appear, however, that this species has been identified in this country with certainty more than twice, although it may possibly have occurred oftener. A specimen shot at Plumpton Bosthill, near Brighton, in September, 1854, was recorded by Mr. Borrer in the "Zoologist" for that year (p. 4511), and was figured by Mr. Yarrell in the third edition of his "History of British Birds" (i. p. 314). A second, obtained at Start Point, Devonshire, in September, 1859, was noticed by Mr. Llewellyn in the "Annals and Magazine of Nat. History," 1859 (iv. p. 399), and in the "Ibis," 1860 (p. 103). It is possible that this may be the Red-tailed Warbler (*Sylvia erythaca*), six specimens of which are stated to have been taken near Plymouth, and to have occurred there for the first time in Britain.[1] From a want of acquaintance with its habits, this bird has been erroneously called the Rufous *Sedge* Warbler. It is never found in the neighbourhood of sedge, but on the driest ground, amidst scrub and thick underwood. In fact, as regards structure and habits, it differs in so many respects from the river warblers that it has been generally separated from them, and, except for convenience, ought not to be included in the present sketch. Its real home seems to be North Africa and Palestine; but it is not uncommon in some parts of Southern Europe, and is found (accidentally only) as far north as the British Islands.

---

[1] *Cf.* Bellamy's "Nat. Hist. South Devon," p. 205.

THE PIED WAGTAIL.

(Motacilla Yarrelli.)

BY many writers on ornithology, the Pied Wagtail has been regarded as a resident species in Great Britain, since it is to be met with in some parts of the country all the year round, but there can be no doubt that large numbers migrate southward for the winter, and return to our shores again in spring. On several occasions when crossing by steamer to the opposite coasts of France and Belgium, I have seen Pied Wagtails passing across and at times even alighting on board the vessel for a short rest.

On quitting the ship they would fly round and

round for some seconds with their peculiar undulatory flight, and finally make off for the land in a straight line, often directly in the vessel's course.

According to the observations of Mr. Knox, the Pied Wagtails which have wintered abroad reach the coast of Sussex about the middle of March, and on fine days may be seen approaching the shore, aided by a gentle breeze from the south, their well-known call-note being distinctly audible from the sea long before the birds come in sight.

The neighbouring fields, where but a short time previously not a bird of the kind was to be seen, are soon tenanted by numbers, and for several days they continue dropping on the beach in small parties. The old males come first, while the females and males of the previous year do not appear until some days later. After resting near the coast for a few days the new comers proceed inland, and any good observer there stationed may perceive how much the numbers of the species increase at this season.

About the middle of August there is a general return movement towards the coast, and the Wagtails now first become gregarious. At that time Mr. Knox has frequently observed them in the interior of the county, where they remain but a few days, making way for fresh detachments, which in their turn follow the same route to the sea. At the end of the month, or early in September, they may be seen of a morning, flying invariably from west to east, parallel to the shore, but following each other in constant succession.

These flights continue from daybreak until about ten o'clock in the forenoon, and so steadily do the birds pursue their course that even when one or more of an advancing party have been shot, the remainder do not fly in a different direction, but opening to right and left close their ranks and continue their progress as before. During this transit their proximity to the coast depends to some degree on the character of the country lying between the South Downs and the sea; but as they advance towards Brighton, the migrating bands, consisting chiefly of the young of the year, accumulate in vast flocks, and thus they seek the adjoining county of Kent, whence the voyage to the continent may be performed with ease and security even by birds but a few months old, and unequal to protracted flights.[1]

The habits of the Pied Wagtail are so generally known, that little need be said here upon the subject. Its partiality for shallow water, where it preys upon aquatic insects, and even small fish, such as minnows and sticklebacks, has led to its being familiarly known as the Water Wagtail, although it is not more aquatic in its habits than other members of the genus, indeed, scarcely so much as one species, the Grey Wagtail, whose haunts seem inseparable from the water-side.

---

[1] For this abstract of Mr. Knox's observations, taken from his "Ornithological Rambles in Sussex," I am indebted to Professor Newton, who has thus ably condensed them in his new edition of Yarrell's "History of British Birds."

THE WHITE WAGTAIL.

(*Motacilla alba.*)

CLOSELY resembling the last-named in form and general appearance, the White Wagtail long escaped observation as an annual summer migrant to this country. Its distinctive characters, however, are now almost universally admitted, and ornithologists experience little difficulty in recognizing the two species.

The particular respects in which the White Wagtail differs from its congeners are noticeable chiefly in the summer, or breeding plumage, when the former has a black cap clearly defined against a grey back, while in the latter the black colour of the head merges in the black of the dorsal plumage and no such cap is discernible. In summer both species have the chin black, and in winter the same parts in both are white. In the immature and winter dress it is not so easy to distinguish them, and in form and structure at all ages and seasons no real difference seems to exist. This has naturally raised some doubt in the minds of many as to the validity of the so-called species, a doubt which is strengthened by the circumstance that in regard to haunts and habits the two may be said to be inseparable.

This much, however, seems to be certain, that whereas the Pied Wagtail is generally distributed as a resident species, migrating southward at the approach of winter, the White Wagtail spends only the summer months in this country, and is then very local in its distribution.

Beyond the British Islands the White Wagtail has a much more extensive range than its congeners, being found throughout the whole of Europe, penetrating to the North Cape and even to Iceland, and travelling southward beyond the Mediterranean into Africa, to within a few degrees of the equator.

THE GREY WAGTAIL.

(Motacilla sulphurea.)

EXCEPT for the purpose of a momentary comparison, it would be beyond the scope of the present volume to notice the Grey Wagtail here, for this bird does not come under the definition of a Summer Migrant.

It is rather a winter visitor, being most frequently observed in the cold season, although many pairs remain in suitable localities throughout the country to nest and rear their young.

Upon this point Professor Newton has remarked that "a line drawn across England from the Start Point, slightly curving round the Derbyshire hills, and ending at the mouth of the Tees, will, it is believed, mark off the habitual breeding-range of this species in the United Kingdom; for southward and eastward of such a line it never, or only occasionally breeds, while to the westward and northward its nest may be looked for in any place suited to its predilections, as above described, whether in this island or in Ireland, where, according to Thompson, it is extensively, though not universally distributed. In Scotland, says Macgillivray, it is rare to the north of Inverness, but it is an occasional summer visitor to Orkney, and in Shetland it occurs towards the end of summer, though it is not known to have been met with in the Outer Hebrides. In the south-west of England its numbers are in summer comparatively small, but it breeds annually in Cornwall and on Dartmoor; and as we pass northward its numbers increase, until in parts of Scotland, perhaps, they attain their maximum. Nests have been reported from Dorset, Wilts, Hampshire, Sussex, and even Kent; but in those counties they are confessedly casual, and only in the case at Chenies, in Buckinghamshire, mentioned by Mr. Gould (" Contr. Orn." 1849, p. 137), does the species seem to have been more than an accidental settler."

The Grey Wagtail may be at once distinguished by having the vent and upper tail coverts of a sulphur-yellow, and by its great length of tail. In summer it has a black patch upon the throat, of a triangular shape when viewed in profile, and bordered with white, but in winter this black patch disappears, and the throat is then of a pale yellowish-white.

It has been stated by Temminck and other naturalists who have followed him, that the black throat is the peculiar attribute of the male bird in the summer or breeding plumage; but this is a mistake. Both sexes have a black throat in the breeding season, as I know from having observed them when paired, and from having examined numerous specimens of which the sex had been carefully ascertained by dissection.

The haunts of the Grey Wagtail are somewhat different to those of its congeners. It affects pools and streams, especially where there is a good current, and may frequently be seen perched upon boulders and mill-dams, where it feeds upon the freshwater limpets (*Ancylus fluviatilis*), and other small mollusca which are found attached in such situations.

The nest is generally placed not far from the water, in some inequality of the bank, or crevice of an overhanging rock. Upon a rugged mountain stream in Northumberland some years since, I daily observed a pair of these birds, and derived much pleasure in watching their building operations. It was some time before I could discover the nest, so skilfully was it concealed, for the birds had selected a crevice in a rock which was much overgrown with moss, and by constructing their nest entirely of this moss, it would easily have escaped observation, had I not patiently watched for the ingress and egress of the owners.

The geographical range of the Grey Wagtail beyond the British Islands has not been satisfactorily determined, in consequence of the difficulty of identifying the species amongst other allied forms which are to be met with in the confines of Europe and Asia. It certainly does not go far north in Europe, perhaps not beyond Northern Germany, but southward it is met with in winter in most of the countries bordering the Mediterranean, as well as in North Africa, Madeira, and the Azores.

THE YELLOW WAGTAIL.

(*Motacilla Rayi.*)

BY many authors the Yellow Wagtails have been separated from the Pied Wagtails under the generic term *Budytes*, proposed by Cuvier, not only in consequence of their very different colouration, but also on account of their possessing a longer and more strongly-developed hind claw. The numerous intermediate forms, however, which the researches of modern naturalists have brought to light from various parts of the Old World, have rendered this subdivision less necessary or desirable than it may originally have appeared to be. In outward form, internal structure, and habits, they are all Wagtails, and one generic term for the whole has, at all events, the merit of simplicity.

The Yellow Wagtail, whose plumage in the breeding season equals in brightness that of the Canary, is one of the most attractive of all our summer migrants. When running over the pastures and fields of sprouting wheat, the olive-green colour of the dorsal plumage renders it very inconspicuous, but when perched upon some rail, or clod upon the bare fallow, the bright yellow of the under-parts contrasts vividly with the duller surroundings, and at once attracts the attention of the passer-by. Its favourite haunts are the marshes and water-meadows where cattle are pastured. Here it finds plenty of food amongst the insects which are disturbed by the grazing kine, and the numerous small and thin-shelled mollusca which abound in such situations.

When the nest has to be constructed—and it

is always upon the ground—more sheltered spots are selected, such as a tussock of rough grass, or the foot of a bunch of tares or clover, and I have occasionally discovered a nest under an overhanging clod upon a bare fallow. Thus in regard to its mode of nesting it differs essentially from the well-known Pied Wagtail. Its note, too, is very different, and its flight much sharper, and with bolder curves. The eggs are quite dissimilar, being so closely freckled over with yellowish-clay colour, like those of the Grey Wagtail, as to appear at a little distance almost uniformly so coloured; whereas the eggs of the Pied and White Wagtails are white, freckled with ash-grey, chiefly at the larger end.

The Yellow Wagtail generally arrives in this country during the first week of April (for many years I have noted the 5th of that month as the average date for its appearance), and it departs during the first week of September. For some time previous to its departure, the young and old assemble in flocks, and it is not unusual to see several united family parties in the meadows, numbering from a dozen to a score of individuals.

Although generally distributed during the summer months throughout the greater part of England and Scotland, it is said to be somewhat rare in Ireland, where its presence has been detected by comparatively few observers. So much more attention, however, is paid to ornithology now-a-days, that this species, like many others, may be reported to be more common than formerly because more observed. In the central and southern portions of Europe it is not uncommon, and crossing the Mediterranean, as winter approaches, it passes down both the east and west coasts of Africa as far as Natal on the one side and Angola on the other. A considerable number, however, pass the winter in Africa, a good many degrees further north.

THE GREY-HEADED WAGTAIL

(*Motacilla flava.*)

SIMILAR in form and general colouration to the last-named, amongst the flocks of Yellow Wagtails that visit us in the spring the grey-headed species no doubt often escapes observation. But it is not on this account to be considered rare. On the contrary, there is good reason to believe that it is a regular migrant to this country, and this is not surprising when we consider that it is the common Yellow Wagtail of northern Europe, the true *Motacilla flava* of Linnæus. It differs chiefly from Ray's Wagtail in having a well-defined cap of a grey colour on the head, a white instead of a yellow streak over the eye, and a white chin instead of a yellow one.[1] It frequents the same situations as the last-named, and its habits are very similar.

---

[1] For some further points of distinction the reader may be referred to "The Birds of Middlesex," pp. 64, 65.

The specimens which have been obtained and recorded as British, and which amount to a considerable number, have been for the most part met with on the coasts of the eastern, southern, and south-western counties of England, and almost invariably in the spring of the year. There can be no doubt that it breeds here; indeed, the fact of its having done so in two or three instances has been already recorded. In the " Zoologist" for 1870 (p. 2343), Mr. J. Watson of Gateshead, near Newcastle-on-Tyne, writes :—" I have seen a good many notices in the 'Zoologist' of the occurrence of the Grey-headed Wagtail : it may interest you to hear of its breeding in this neighbourhood. Two nests were found by a friend of mine last year on some swampy ground near here. This year on the 13th of June I found another ; and on the 8th of July my friend shot two young birds beginning to assume their mature plumage : one of these birds is in the possession of and was identified by Mr. John Hancock of Newcastle."

But although the greater number of recorded British specimens have been obtained in the South of England, a few have been noticed from time to time in Scotland, and Dr. Saxby has on several occasions seen the species even as far north as Shetland. Mr. Blake Knox thinks that it occurs in Ireland, but that it is probably much overlooked, or perhaps confounded with the last-mentioned species. As it is common in summer in most of the countries of Western Europe, one would naturally expect to meet with it more frequently at the same season in Great Britain; and the increasing attention which is being paid to ornithology, and especially to the birds of particular districts, will no doubt result in the establishment of this species in the list of British birds as an annual summer migrant.

THE MEADOW PIPIT.

(Anthus pratensis.)

PREMISING that attention is not confined to species which are British, it is generally admitted by ornithologists that the Pipits are a difficult group to identify. They are subject to such variation in size and colour that it has often happened that one and the same species has been described four or five times as new, under as many new names. Gradually, however, as the researches of naturalists become extended, and the transport of specimens from various quarters of the globe is facilitated, the difficulty wears off, and we are enabled to define with sufficient accuracy the limits of each species and the variations of plumage within those limits.

Were I to confine my remarks in the present instance to those Pipits only which are regular summer migrants to this country, I should not have to mention more than two species. It may be well, however, to take a glance at all those which have a claim to be included in the British list, distinguishing them under the heads of "Residents," "Summer Migrants," and "Occasional Visitants."

Two species only are resident with us throughout the year—the well-known Meadow Pipit or Titlark (*Anthus pratensis*), and the larger Rock Pipit (*Anthus obscurus*). Both these, however, are to a certain extent migratory at the approach of winter, assembling in small flocks, and moving from place to place in search of food. The Tree Pipit (*Anthus arboreus*) visits us regularly in April, and remains in this country until September; and there can be little doubt, from recent observations of naturalists in different parts of the country, that the Water Pipit (*Anthus spinoletta*, Linnæus, or *Anthus aquaticus*, Bechstein) is also an annual summer migrant to our shores. At irregular intervals, and in addition to these, we are occasionally visited by Richards' Pipit, the Tawny Pipit, the Red-throated Pipit, and the Pennsylvanian Pipit. Of the two resident species, as well as the Tree Pipit, it can scarcely be necessary to say much, for their appearance and habits, if not well known to all, are described in almost every book on British birds. After pointing out their distinguishing characters, therefore, my remarks will refer chiefly to the geographical distribution of the species.

The Pipits hold an intermediate place between the Wagtails and Larks, having the slender bill of the former, and, with one exception, the long hind claw of the latter. Like these birds, they live almost entirely on the ground, where they seek their food, build their nests, and rear their young. Low-lying meadows and marshy places, the margin of tidal harbours, and the seashore are the favourite haunts of the Pipits. In such situations, except in very hard weather, they find abundance of food, consisting chiefly of insect larvæ, small beetles, flies, seeds, and minute univalve mollusca. I have almost invariably found, in addition, that the stomachs contain little particles of grit or brick, swallowed no doubt to assist in triturating the food.

The Meadow Pipit (*Anthus pratensis*) is the smallest as well as the commonest species to be met with, and is generally dispersed throughout the British Islands, including Orkney and Shetland. It is by no means confined to the plains or open country, but is frequently to be met with on mountain sides, sometimes at a considerable elevation. Tourists and sportsmen must doubtless have remarked this when climbing the Scotch and Irish mountains. The late Mr. Wheelwright, in Lapland, found it "very high up on the fells;" Professor Salvadori remarked it on the Apennines; and Messrs. Elwes and Buckley include it in their list of the birds of Turkey as frequenting the mountains.

In summer it is common in Scandinavia, and Mr. Wheelwright found it nesting in Lapland. It goes as far north as the Faroe Isles and Iceland.[1] According to Professor Reinhardt,[2] Dr. Paulsen, in Sleswick, received a single specimen from Greenland in 1845; but he adds that he (Professor R.) never saw it there himself. The Meadow Pipit appears to be generally distributed throughout Europe, and at the approach of winter emigrates in a south-easterly direction by way of Sicily and the Ionian Islands to Palestine. Lord Lilford states that it is very common in Corfu and Epirus in winter.[3] Canon Tristram found it in large flocks throughout the winter in North Africa, "apparently on passage;" and in Southern Palestine and in the Plains of Sharon he remarked that it was very abundant. According to Sir R. Schomburgk, it occurs as far eastward as Siam; but Mr. Blyth considered the Siamese *pratensis*

---

[1] See Professor Newton's remarks on "The Ornithology of Iceland," appended to Baring Gould's "Iceland; its Scenes and Sagas," p. 409.

[2] "Ibis," 1861, p. 6.   [3] "Ibis," 1860, p. 229.

to be the Red-throated Pipit (*A. cervinus*) in winter plumage. It is known to occur in India, however, as Mr. Hume has procured this species near Ferozpore, North-west India; and Mr. Blyth saw specimens from other parts of the Northwest provinces. The range of this bird southwards, that is through Africa, seems to be very limited. According to Mr. Saunders, it is common in Spain in winter, but it is not included in Mr. Tyrwhitt Drake's list of the birds of Morocco; and though Mr. Salvin shot a specimen at Kef Laks in the Eastern Atlas, it appears to occur in North-west Africa exceptionally. The Pipit of the Canaries, originally regarded as *A. pratensis*, has been described by Dr. Bolle[1] as distinct, under the name of Berthelot's Pipit (*Anthus Berthelotii*). But Mr. Vernon Harcourt maintains—and so did the late Mr. Yarrell—that Madeiran specimens can in no degree be distinguished from specimens of *A. pratensis* from other parts.

---

[1] "Ibis," 1862, pp. 343, 348; and "Journ. f. Orn." 1862, pp. 357, 360.

THE ROCK PIPIT.

(*Anthus obscurus.*)

THIS Pipit, as already observed, is to be found on most parts of our coast throughout the year, except on that portion which extends from the Thames to the Humber, where it is only observed in spring and autumn during the period of migration. For although a resident species, inasmuch as individuals may be found on some parts of the coast throughout the year, it is also, to a certain extent, migratory, receiving a considerable accession to its numbers in spring, and a corresponding diminution in autumn. It may be distinguished from the common Meadow Pipit by its larger size, longer bill, tarsus, and toes, and by its having the upper portion of its plumage of a greener olive. The legs are of a much darker brown, and I have remarked that in freshly-killed specimens the soles of the feet are yellow, a circumstance which appears to have been generally overlooked, but which is worth noticing as an addition to its distinguishing characters. A considerable difference also will be observed in the two outer tail feathers on each side. In the Meadow Pipit the outermost tail feather is for the greater part white, and the next has half the tip of the inner web also white. In the Rock Pipit the same parts of these feathers are not white, although conspicuously lighter than the remaining portion.

The Rock Pipit found in Scandinavia (*Anthus rupestris* of Nilsson), is considered by some to be distinct from the species which frequents our own shores, but, as I think, on extremely slender grounds. The points of difference have been thus stated: "They consist, so far as we can ascertain, merely in the presence of a bright buff or pale cinnamon tinge on the breast of the male in *A. rupestris*, and perhaps in that form being of a slighter build than *A. obscurus*. In the female of the so-called *A. rupestris* the warm colour is much more faintly indicated; in some specimens it is doubtful whether it exists at all. The outer tail feathers, which in *A. spinoletta* afford so sure a diagnosis, are in *A. rupestris* just as dingy as in *A. obscurus*."

There can be no doubt that the chemical constituents of colour in the plumage of birds are always more or less affected by climatic agency; and, this being so, one can hardly be justified in founding a new species on mere variation of colour, where there is at the same time no modification of structure. There can be little doubt that the Scandinavian Rock Pipit is identical with our own bird, the slight differences observable being easily accounted for through climate and the season of the year at which specimens are obtained.

The late Mr. Wheelwright makes no mention of this bird when treating of the ornithology of Lapland. Messrs. Godman met with it on the seashore at Bodö, Norway, " in tolerable abundance," and Mr. Hewitson also saw it in Norway. Although Temminck says that it goes as far north as Greenland, this does not appear to be the case; for Professor Reinhardt, who has paid especial attention to the ornithology of Greenland, states that only two species of Pipit are to be met with there—namely, the American *Anthus ludovicianus*, which breeds there, and *A. pratensis*, of which, as above stated, a single specimen only is recorded to have been obtained. It is rather remarkable that Professor Blasius has not included the Rock Pipit in the avifauna of Heligoland, seeing that *A. cervinus*, *A. ludovicianus*, and *A. Richardi* are all stated to have been taken on that island.[1]

Although found upon the shores of Holland,

---

[1] *Cf.* "Naumannia," 1858, p. 425, and "Ibis," 1862, p. 71.

Belgium, and France, it either goes no farther to the south-west, or else it has been overlooked; for neither Mr. Howard Saunders, in his "List of the Birds of Southern Spain," nor the Rev. A. C. Smith, in his "Sketch of the Birds of Portugal," give it a place in the avifauna of those countries. Mr. C. A. Wright states ("Ibis," 1869, p. 246) that he has only obtained a single specimen in Malta. Further eastward, namely, on the coasts of Epirus and Corfu, Lord Lilford found it to be common, and on this account it has been included by Messrs. Elwes and Buckley in their "List of the Birds of Turkey." I am not sure whether it has been met with in Asia Minor, but probably it does not extend either eastward or southward beyond the coast line of the Mediterranean. The observations of naturalists certainly tend to prove that its proper habitat is Northern Europe, and perhaps nowhere is it commoner than in the British Islands.

THE TREE PIPIT.

(*Anthus arboreus.*)

ALTHOUGH a regular summer visitant to England, the Tree Pipit, like the Nightingale, from some unexplained cause, is distributed over a very limited area. It never reaches Ireland, and is considered rare in Scotland, although the nest has been found as far north as Dumbarton, Aberdeen, Banff, and East Inverness.[1] Even in Wales and Cornwall it is a scarce bird, so that England may be said to be the

---

[1] *Cf.* A. G. More, in the "Ibis," 1865, p. 123.

western limit of its geographical range. Mr. Wheelwright never met with it in Lapland, but Messrs. Godman found it in June as far north as Bodö, in Norway, and from this latitude southwards to the Mediterranean it seems to be well known in summer. Mr. Howard Saunders says that it is generally distributed in Spain from autumn to spring, and he suspects that some remain to breed on the high plateaux. In Portugal, according to the Rev. A. C. Smith, it is rare. Mr. Wright, of Malta, states that it is very common in the island in spring and autumn, departing in May northwards, and returning in September and October. He adds that a few remain the winter. According to the observations of Lord Lilford, it is now and then seen at Corfu in winter, throughout which season it is found in small flocks, apparently on passage to North Africa. Mr. Layard does not include it in his "Birds of South Africa," but, according to Professor Sundevall ("Svenska Foglarna," p. 41), a specimen was killed by Wahlberg on the Limpopo, in Kaffirland, between lat. 25 deg.

and 26 deg. S. Canon Tristram found it sparingly distributed in Palestine in winter, and in spring in the Jordan valley. It is recognised by naturalists in north-west India, and there can be little doubt that the Pipit which has been described from that country, and from China and Japan, under the name of *Anthus agilis*, Sykes, is only our old friend *A. arboreus* in a different plumage from that which it assumes here in summer. Herr von Pelzeln says [1] that *agilis* only differs from *arboreus* in having a stouter bill, and he does not think that it can be specifically distinct, notwithstanding that Dr. Jerdon gives both species as inhabitants of India. On this point Mr. Hume says ("Ibis," 1870, p. 287): "I took nine specimens of *arboreus* from England and France, and compared them with our Indian birds. There was no single one of them to which an exact duplicate could not be selected from amongst my Indian series. That all our Indian Pipits known as *agilis, maculatus,*

---

[1] *Cf.* "Journ. für Orn.," 1868, pp. 21-37.

and *arboreus* ought to be united as one species under the latter, or possibly some older, name, I can now scarcely doubt."

THE WATER PIPIT.
(*Anthus spinoletta.*)

IN size this bird equals our well-known Rock Pipit, but may be distinguished by the vinous colour of the throat and breast, by the absence of spots or streaks upon the under parts, and by the outer tail feathers, which are marked with white, as in *A. pratensis*. It was named *spinoletta* from the provincial name applied to the bird in Italy, whence Linnæus described it.[1] Pallas, however, altered the name to "*pispoletta*," because Cetti affirmed that this was the correct Florentine term, and not *spinoletta*. Linnæus's name, nevertheless, on the ground of priority, is entitled to precedence. The species was identified with *aquaticus* of Bechstein by Bonaparte.[2]

---

[1] "Syst. Nat.," i. p. 288.   [2] "Consp. Av." i. p. 247.

This bird seems to have been first made known to English naturalists by Mr. Thomas Webster, of Manchester, who, in a communication to the "Zoologist" (p. 1023), stated that he had seen three birds at Fleetwood in October, 1843, which he had not the slightest hesitation in identifying with a Pipit described by M. Deby as *Anthus aquaticus*, Bechstein, and which to all appearance were totally distinct from the common Rock Pipit of our coast. In January, 1860, the Rev. M. A. Mathew, in a letter to Mr. Gould, called attention to the fact of his having procured a Pipit at Torquay, which was subsequently identified unhesitatingly with *A. aquaticus* of Bechstein. Since that date, Mr. Gatcombe, of Plymouth, has noticed several other specimens in Devonshire, and a great many have been procured in Sussex, chiefly in the neighbourhood of Brighton. Thus the claim of this bird to rank as a British species has come to be pretty well established. M. Baily, in his "Ornithologie de la Savoie," says that the Water Pipit is common at all seasons of the year both in Switzerland and Savoy. During winter it frequents the wet meadows, marshes, and unfrozen springs in the valleys, and about the end of March or beginning of April ascends the mountains, and resorts to the most sterile plateaux, fields, heaths, and stony places in the neighbourhood of water, where it nests on the ground under stones, sometimes in clefts in the rock, but oftener in the grass beneath the bilberry, whortleberry, or some creeping bush.

In the fall of the year it descends to the warmer valleys and frequents the margins of the rivers, whence it has derived the name of Water Pipit, making its way gradually southward as winter approaches. Mr. Saunders has met with it at Malaga in winter; but apparently it is not common in Spain, and, according to the Rev. A. C. Smith ("Sketch of the Birds of Portugal") still less so in Portugal. Mr. Wright has met with it once in Malta, having shot a specimen there in November, 1860. It crosses the Mediterranean to North Africa. Canon Tristram met with it in Algeria, and Captain Shelley recognised it in Egypt. In the peninsula of Sinai it was found by Mr. C. W. Wyatt, frequenting the sides of the salt-ponds near Tor, and it is included in Mr. Strickland's list of the birds of Asia Minor ("P. Z. S.," 1836, p. 97) as being found on the coast in winter at Smyrna, whence it penetrates to Palestine (Tristram, "Ibis," 1866, p. 289). Messrs. Elwes and Buckley have enumerated this amongst other species in their list of the birds of Turkey, and Ménétries states ("Cat. Rais. Caucas." p. 39) that it is common on the shores of the Caspian in April, May, and June. The range of this bird eastward is at present hardly determined; partly, perhaps, because the Pipits have been a good deal neglected for the sake of more attractive species, and partly on account of the difficulty which travellers usually experience in the identification of this difficult group of birds. That the Water Pipit penetrates to north-west India is to be inferred from the fact that Mr. Hume sent M. Jules Verreaux a specimen for identification from the Punjab west of the Sutlej.

RICHARD'S PIPIT.

(*Anthus Richardi.*)

OUT of compliment to the zealous amateur who first made known an example captured in autumn in Lorraine, the name of Richard's Pipit has been bestowed on this bird, which is becoming better known to ornithologists in this country every year. Its superior size, stouter bill, greater length of leg, and longer hind claw, at once serve to distinguish it from the commoner species. As compared with the Rock Pipit, the largest of those with which we are most familiar, its dimensions are as follows:

	Bill.	Wing.	Tarsus.	Hind toe with claw.
	Inches.	Inches.	Inches.	Inches.
A. obscurus	·5	3·2	·9	·8
A. Richardi	·6	3·6	1·2	1·2

Its occurrence in England has been noted, as might be expected, chiefly on the east and south coasts, in every month between September and April, both inclusive. At least fifty specimens have been seen or procured, distributed as follows: Northumberland, 2; Norfolk, 5; Shropshire, 1; Oxford, 1; Middlesex, 12; Kent, 3; Sussex, 5; Devonshire, 11; Cornwall and Scilly, 8. In the west of England, therefore, it would appear to be very rare, and in Ireland it is unknown.

The most northern locality, I believe, whence this species has been procured, is Heligoland, on which island, according to Professor Blasius, it is said to have been obtained by Herr Gätke.[1]

---

[1] *Cf.* "Naumannia," 1858, p. 425.

When staying at Antwerp in May, 1870, I saw three or four specimens which had been taken in that neighbourhood, but the owner of them considered the bird a rarity there. Mr. Howard Saunders obtained a couple near Malaga in the month of February, and learnt that in some winters it is not uncommon in southern Spain ("Ibis," 1870, p. 216). Signor Bettoni, in his grand work on the birds which breed in Lombardy, mentions Richard's Pipit as one of the characteristic species of the Lombard plains. "Nevertheless," says Mr. Saunders ("Ibis," 1869, p. 392), "he must not be understood to mean that it is in any way abundant, or even constant in that province; for the Count Turati assured me that it has never been discovered breeding there, and that, judging from the number of specimens enumerated as obtained in England, it is more common with us than with them. That its appearance is confined to the plains of Lombardy is probably the author's meaning." In Malta it is only found accidentally in spring and autumn, and

Mr. Wright, who has paid so much attention to the ornithology of that island, has only been able to mention three examples as having come under his own notice.

It is rather singular that this bird should not cross the Mediterranean, and be found with other European Pipits during the winter months in North Africa. Nevertheless, I have not been able to find any mention of it in any of the North African lists which I have consulted, neither is it included in the late Mr. Strickland's List of the Birds found in Asia Minor in winter ("P. Z. S.," 1836, p. 97).

It is much commoner, however, in Asia than in Europe. Mr. Hodgson found it in Nepal,[1] and Mr. Hume says it breeds in Ladakh; Mr. Blyth has recorded its occurrence in the neighbourhood of Calcutta, and Mr. Blanford met with it in the Irawadi Valley. It is included by Sir R. Schomburgk in his List of the Birds of

---

[1] Capt. Beavan recorded it from Simla ("Ibis," 1868, p. 79), but Mr. Hume showed this to be an error, the species mistaken for it being *A. sordida* ("Ibis," 1869, p. 120).

Siam ("Ibis," 1864, p. 249), and, according to Mr. Swinhoe, is common in North China (Takoo and Peking) in September, and in Amoy, Formosa, and Hainan in winter.

THE TAWNY PIPIT.

(*Anthus campestris.*)

EASILY mistaken for Richard's Pipit, this bird is, however, of a more sandy colour, and may be distinguished by its short hind claw. In Richard's Pipit, it will be remembered, the hind claw is very long. Its real habitat may be said to be North Africa and Palestine. Canon Tristram calls it the common Pipit of the Sahara, and Mr. O. Salvin found it abundant on the plateau of Kef Laks and on the plains of Djendeli, in the Eastern Atlas. In Upper Egypt and Sinai it is occasionally plentiful, and is found all over the cultivated coast and hill districts of Palestine, where it is a permanent resident.

"The soil of the Sahara," says Mr. J. H. Gurney, jun. ("Ibis," 1871, p. 85), "is in some places soft and sandy, in others hard and pebbly. The Tawny Pipit affects the former, where there is little or no herbage. Its flight is undulating, like that of the Wagtails; and, like the latter, it twitters on the wing." Canon Tristram, referring to the habits of this species in Palestine, where he obtained several nests on the bare hills, says ("Ibis," 1866, p. 289), "It is one of the tamest of birds, and particularly affects the mule paths, flitting along in front of the traveller, and keeping unconcernedly a few yards ahead." "The nest," says Mr. Salvin, "is composed of roots, with a lining of horsehair, and is placed on the lee side of a bush. The eggs vary very much, some being light-coloured, and almost like wagtails', while others are much darker and more profusely marked."

Although, as above stated, North Africa and Palestine may be regarded as its home, the Tawny Pipit ranges a long way to the north and south of this tract, and is common in some parts of Southern Europe in summer. It is found as far northward as Sweden—where, as Mr. Wheelwright has remarked, it is confined to the sandy shores of the south—and accidentally in England, where specimens have been several times procured on the coasts of Sussex, and in Cornwall.[1]

Lord Lilford has observed that it is common in Spain in summer ("Ibis," 1866, p. 178), an observation more recently confirmed by Mr. Howard Saunders ("Ibis," 1869, p. 392). In Portugal, according to the Rev. A. C. Smith, it seems to be equally well known.

It is annually observed in Malta in spring and autumn, but never found there during the winter months (Wright, "Ibis," 1864, p. 61). Lieut. Sperling, however, believes that it is not uncommon on the north coast of the Mediterranean in winter. South of the habitat assigned to it, this bird ranges through Abys-

---

[1] *Cf.* Dawson Rowley, "Ibis," 1863, p. 37, and 1865, p. 113; Bond, "Zoologist," 1870, pp. 1984 and 2383; and Rodd, "Zoologist," 1868, p. 1458.

sinia (whence I have seen a specimen in the collection of African birds belonging to Mr. Sharpe) to Mozambique, where, according to Lieut. Sperling, it is plentiful in winter; and Mr. Layard has included it amongst the birds of South Africa, having received specimens from Windvogelberg and the Knysna. It has a West African representative in *Anthus Gouldii* of Frazer (Hartlaub, "Orn. West Afr.," p. 73), which differs in its smaller size and darker colour, and in having the head of a uniform dull brown, instead of being streaked.

THE PENNSYLVANIAN PIPIT.

(*Anthus ludovicianus.*)

ON the authority of several good naturalists this species is stated to have occurred several times in the British Islands; but the general description of the specimens referred to applies as a rule so well to the *Anthus spinoletta* above mentioned, that it is extremely difficult to say to which of the two species they belonged. It is of course far more probable that the visitors to our shores would be of European, not American, extraction. At the same time they have been described as according so well in every respect with the American *ludovicianus*, that we must either admit that the latter bird occasionally visits this country, or agree with Richardson and Swainson ("Faun. Bor. Americana," ii. p. 231) that it is indistinguishable from *aquaticus* of Bechstein, that is, *spinoletta* of Linnæus.

Edwards was the first to notice this bird as a visitant to England, giving a description and figure of a specimen obtained near London in his "Gleanings" (vol. ii. p. 185, pl. 297). Montagu shortly afterwards noticed two in his "Ornithological Dictionary," one of which had been taken in Middlesex, the other near Woolwich.

Macgillivray, in his "Manual of British Birds," p. 169, minutely describes two Pipits which were shot near Edinburgh in June, 1824, and which he identifies clearly with the American species.

Mr. Turnbull, in his " Birds of East Lothian," states (p. 40) that three Pennsylvanian Pipits were shot at Dunbar in East Lothian by Mr. Robert Gray, of Glasgow.

Mr. Bond has a Pipit, identified as belonging to this species, which was obtained at Freshwater, in the Isle of Wight, in September, 1865; while the most recent instance of the occurrence of this Pipit in England will be found in the " Zoologist" for 1870. But anyone who reads the correspondence relating to this instance ("Zool." tom. cit. pp. 2021, 2067, and 2100) will see how difficult it is to identify a species when the specimen is not in fully adult plumage.

When it is remembered that *Anthus ludovicianus*, as stated by Professor Reinhardt (" Ibis," 1861, p. 3), breeds in Greenland, and, according to Professor Blasius, is found in Heligoland ("Naumannia," 1858), it is certainly not improbable that it should occasionally be found in the British Islands. At the same time it is very desirable that some more convincing evidence than that which already exists of its occurrence here should be placed upon record.

THE RED-THROATED PIPIT.

(*Anthus cervinus.*)

THE present bird has, as yet, been scarcely admitted into the British list. I have seen a specimen in the collection of Mr. Bond, which was killed at Unst, Shetland, on the 4th May, 1854, and about the same year, but in September, another in the same collection was shot at Freshwater in the Isle of Wight.

In the adult plumage the species is easily recognized by the ruddy brown colouring of the upper portions of the plumage, and by the rufous patch upon the throat.

In size it is equal to the Meadow Pipit, and by some naturalists it has been considered a permanent race or variety of that species; but the observations of Prof. Newton on this point[1] certainly tend to show that the species is a valid one. It was met with by him in June, 1855, when in company with Messrs. Wolley and Simpson, in a restricted locality in East Finmark, between Wadsö and Nyborg, and several well-identified nests were procured. A specimen procured in Heligoland is in Herr Gätke's collection.

It is not uncommon as a winter visitant in Turkey, and Mr. Wright has shot many specimens in Malta, where he says it arrives in small flocks in spring and autumn. In Egypt and Nubia this bird quite takes the place of *A. pratensis*, and is sometimes very common there. It probably winters also in Palestine, although Canon Tristram, during his sojourn there at that season, only met with a single specimen on the coast of the plain of Sharon. It has been found in China, Japan, Formosa, and Hainan, by Mr. Swinhoe, who suggests that this bird

---

[1] See Bree's "Birds of Europe," vol. ii. p. 155.

in its winter plumage is the *Anthus japonicus* of Temminck and Schlegel. Mr. Blyth thinks that it should probably be erased from the Indian list, as the ordinary Himalayan species, *A. rosaceus* of Hodgson, has been confounded with it. Upon this point, however, much difference of opinion prevails. Dr. Jerdon, in his "Birds of India," gives *rosaceus* as a synonym of *cervinus*, and Mr. Hume is puzzled to distinguish *rosaceus* from *arboreus*. He says ("Ibis," 1870, p. 288): "Typical examples of both species seem unmistakably distinct, but intermediate forms of the most puzzling character occur, of such a nature that it really seems to me impossible to decide to which species they ought to be referred."

Professor Newton considers that the Red-throated Pipit is as yet scarcely entitled to a place in the list of British Birds; nevertheless it is a bird, as he says, whose migratory habits and wide north-eastern range make it very likely to occur in this country, and probably its recognition as an occasional visitor to the British Islands is only a matter of time and observation.

THE SPOTTED FLYCATCHER.

(*Muscicapa grisola.*)

THE family of Flycatchers is a very large one, having representatives in all parts of the globe; but in the British Islands two species only can with propriety be included in the list of annual summer migrants. It is true that at least one other species has been met with in this country, to which allusion will be made presently; but it cannot be regarded in any other light than that of a rare and accidental visitant.

The Spotted Flycatcher (*Muscicapa grisola*),

as remarked by the eminent Irish naturalist, Thompson, is probably little known, except to the observant ornithologist. Owing to the dulness of its plumage, its want of song, and its weak call being seldom heard, it is certainly one of the least obtrusive of our birds; the trees, too, having put forth their "leafy honours" before the period of its arrival, further serve to screen it from observation. It is one of the latest of our summer migrants to arrive, seldom appearing before the second week in May, and generally taking its departure during the first week of September. It is found throughout the British Islands, but is much less common in Scotland. It has, however, been found breeding as far north as Sutherland and Caithness. The situation selected by this bird for its abode during its stay with us is generally in the neighbourhood of gardens and orchards, where it takes up its quarters on a wall or fruit tree, and sallies forth into the air after passing insects. The name of Spotted Flycatcher is more appropriately bestowed upon the bird in its immature plumage, when each brown feather is tipped with a buff spot. As it grows older, these spots gradually disappear. It is a wonderfully silent bird, and even when the hen is sitting the male does not, like the males of so many other species, pour forth a song to enliven her. The nest is usually placed on a beam in a shed, in a hole in a wall, or on the branch of a wall-fruit tree, partially supported by the wall; not unfrequently it may be discovered in a summer-house. It is neatly composed of moss and fine roots, and lined with grass, horsehair, and feathers. The eggs, generally five in number, are bluish white, spotted, chiefly at the large end, with reddish brown.

The late Mr. Wheelwright found the Spotted Flycatcher inhabiting Lapland in summer, but observed that it was not nearly so common there at that season as the Pied Flycatcher. In Central and Southern Europe it is a summer resident, passing through Spain and Portugal, Italy, Turkey, and the Ionian Islands twice a year—namely, in spring and autumn. Its course in autumn appears to be south-east by south. Mr. Wright has noticed it as very common in spring and autumn in Malta, arriving there somewhat later than the Pied Flycatcher. It has been noticed by Mr. J. H. Gurney, jun., as plentiful in Algeria in summer. Captain Shelley met with it once at Alexandria in May, when it was probably migrating; and Rüppell includes it without hesitation amongst the birds of North Africa.[1] In the middle of October, Von Heuglin found that it was not rare near Tadjura, and somewhat later in the year on the Somali coast. In Palestine Canon Tristram found it breeding in all parts of the country, its favourite nesting-places being in the branches of old gnarled trees overhanging the paths ("Ibis," 1867, p. 361). How far eastward it extends I am not sure, as in China and Japan an allied species appears to take its place. But south of the Mediterranean it penetrates to South Africa. Mr. Layard says,[2] "the common European flycatcher has been brought by Mr. Andersson from

---

[1] "Syst. Uebers. d. Vögel N.-O. Afrika's," p. 61.
[2] "Birds of South Africa," p. 148.

Damara Land in some abundance. And Andersson himself states [1] that the bird is common in Damara and Great Namaqua Land, and is found there throughout the year. Dr. Hartlaub cites it on M. Verreaux's authority as from the Cape, and Swainson also alludes to it as from South Africa. Since the publication of the work above quoted, Mr. Layard has been enabled to add that his son procured this bird at Grootevadersbosch, near Swellendam. From Lapland, then, to the Cape of Good Hope, and from Portugal to Palestine is a pretty extensive range for so small and weak a bird as our Common Flycatcher. I should not be surprised to hear that it is found even still further to the eastward, for so many of our summer migratory birds spend their winter in India and China, and after all the greater part of their journey would be by overland route, which admits of their travelling by stages, to rest and feed by the way.

---

[1] "Notes on the Birds of Damara Land," by the late C. J. Andersson; arranged and edited by J. H. Gurney, 1872, p. 129.

THE PIED FLYCATCHER.

(Muscicapa atricapilla.)

FROM its conspicuous black and white plumage, the Pied Flycatcher is a much more attractive species than the commoner bird. Strange to say, although of similar habits, and living on similar food, it is by no means so common as a species, nor so generally dispersed. Its presence in Scotland is always looked upon as an uncommon occurrence, and in Ireland, until recently, it was quite unknown.

During the month of April, 1875, Mr. Robert Warren, jun., of Moyview, Ballina, co. Mayo, met with this bird for the first time in his neighbourhood, and the following communication from him on the subject was published in the natural history columns of "The Field," on the 1st of May, 1875:—"It may interest some of your ornithological readers to learn that a Pied Flycatcher (*Muscicapa atricapilla*) visited this extreme western locality on the 18th of April. My attention was first attracted by seeing it catching insects in the true flycatcher style; but, thinking it rather strange that our common Spotted Flycatcher should appear a month or six weeks earlier than usual, I watched it attentively for some time. It then struck me as having a smaller head and closer plumage than the spotted one, and occasionally I thought I observed some white marks on the wings; but, the evening light just fading, I could not be quite certain of the white marks. Although knowing it to be a flycatcher, I was not satisfied as to its identity, so next morning I returned to that part of my lawn where I had seen it the night before, and again saw it hard at work; but now having better light, and the aid of a field glass, I was not long in making out quite distinctly the white wing marks, which showed me that it was not the common *Muscicapa grisola.* I took my gun and secured what I believe to be the first specimen of *Muscicapa atricapilla* ever shot in Ireland. Neither Thompson in his 'Birds of Ireland,' nor Professor Newton in his new edition of 'Yarrell's British Birds,' mentions it as a visitor to Ireland, or gives any record of its capture in this island; and Mr. Harting, in his 'Handbook of British Birds,' p. 10, says it is unknown in Ireland. The specimen, an adult female, is now in the collection of the Royal Dublin Society."

To this communication the editor appended the following note:—" Although we always regret to hear of the wanton destruction of a rare bird, we must admit that circumstances sometimes occur to justify an individual capture, and we think the present instance is a case in point. By the actual possession of the bird seen, Mr. Warren has been enabled to establish beyond doubt the fact of the occurrence in Ireland of a species previously unknown there, and has thus a complete answer to any sceptic who might suggest that he may have been mistaken in his identification of it."

In England the Pied Flycatcher is a regular summer migrant, quite as much as any other of the small birds already noticed. Mr. A. G. More, in his "Notes on the Distribution of Birds in Great Britain during the Nesting Season," regards it as a very local species, and observes that the nest has occasionally been found in North Devon, Somerset, Dorset, Isle of Wight, Surrey, Oxford, Norfolk, Gloucester, Shropshire, Leicester, and Derby. To these counties I may add Middlesex (for I have known several instances of this bird nesting as near London as at Hampstead, Highgate, and Harrow) and Essex, where the species has been met with at Leytonstone. Yarrell adds Sussex, Suffolk, Yorkshire (where I also have seen it), Worcester, Lancashire, Derbyshire, Cumberland, Westmoreland, Northumberland, and Durham, and on the southern coast, Hampshire. He makes no mention of its occurrence in Wales, neither does Mr. A. G. More in his essay above mentioned. During the summer of 1871, however, several letters appeared in the natural history columns of "The Field," communicating the fact of its nesting in Breconshire, Denbighshire, and Merionethshire.[1] The sites selected for the nests are usually holes in walls, ruins, and pollard trees, and the nest itself is composed of roots, grass, strips of inside bark and horsehair. The eggs, five or six in number, are of a very pale blue colour, much paler, smaller, and rounder than those of the hedge sparrow. A correspondent who has taken several nests of this bird states that he never found one containing feathers; but I think I have seen one lined with feathers which had been taken out of an old birch tree in Lapland by the late Mr. H. Wheelwright. In this lamented naturalist's entertaining book, "A Spring and Summer in

---

[1] See "The Field" for May 27th, June 8th, and June 24th, 1871.

Lapland," he states that, although he never met with the Pied Flycatcher on the fells, it was to be found as far north as the birch region extends, and he generally found the nest in small dead birch stubbs by the riverside. Messrs. Godman met with it some way up the mountains to the north of Bodö in Norway, where the birch was also the favourite nesting tree. As it is common in most parts of Central and Southern Europe, and is found as far westward as Portugal, it is rather curious that Professor Savi should have so long overlooked its occurrence in Tuscany. Dr. Giglioli noticed it as abundant at Pisa in April, and, on recording it as new to the Tuscan avifauna, he added ("Ibis," 1865, p. 56): "When I showed the numerous specimens I had procured to Professor Savi, he was much surprised, and said that, during the forty years he had been studying the ornis of this part of Italy, he had never come across the Pied Flycatcher, which, however, abounds during the spring passage at Genoa, and all along the Riviera." It is a spring and autumn visitor in

Malta; but, though often seen in the valleys and by roadsides in the neighbourhood of trees, it is not so numerous in the island as *M. grisola*. Mr. O. Salvin found the Pied Flycatcher not uncommon about Souk Harras in the Eastern Atlas, and Mr. Tyrrwhitt Drake saw it during the spring migration in Tangier and Eastern Morocco. A specimen from the River Gambia is in the collection of Mr. R. B. Sharpe. Mr. J. H. Gurney, jun., during a recent tour in Algeria, encountered this amongst other familiar birds. He says ("Ibis," 1871, p. 76): "It was not until April that I saw this species, after which it became common. In the dayats and in the Gardaia, where they most abounded, the proportion of adult males in full summer plumage to young birds and females was as one to five. They looked exceedingly picturesque in the rich foliage of the oases, clinging perhaps to a rough palm stem, though their more usual perch was the upper bough of a bush, whence they would dart off after passing flies." To this I may add that the note frequently repeated is not unlike that of the Redstart, although softer and more agreeable, and the bird when uttering it often shuffles its wings after the manner of a Hedge Sparrow. Canon Tristram found this bird to be a summer resident in Palestine, and first noticed it in Galilee on April 23rd; but, though remaining to breed, he considered it rather a scarce bird there.

An allied species, *Muscicapa albicollis*, is generally distributed over the South of Europe, Palestine, and North Africa, which differs from the Pied Flycatcher in having the nape of the neck white instead of black; in other words, the white of the throat extends entirely round the neck. It is found in Greece, Turkey, Tuscany, Spain, Portugal, and France, less commonly in the north of France, and not in Belgium or Holland. It is singular, considering that the two species occupy the same haunts during a great portion of the year, that the White-necked Flycatcher never accompanies its more sable congener to England; yet, so far as I am aware, there is no instance of its occurrence here on record.

What is the cause which operates to restrain one species from migrating, when a closely allied bird of similar habits is impelled to take a long and perilous journey? Truly it is a curious question.

Before taking leave of our British flycatchers, it may be observed that a third species, the Red-breasted Flycatcher (*Muscicapa parva*), a native of South-eastern Europe and Western Asia, has been met with and procured on three separate occasions in Cornwall. One was taken at Constantine, near Falmouth, on Jan. 24, 1863.[1] A second was captured at Scilly in October of the same year;[2] and a third was procured also at Scilly on Nov. 5, 1865.[3] All the specimens procured were immature. The adult bird has a breast like a robin, which renders it a particularly attractive species. It is said to be not un-

---

[1] "Zoologist," 1863, p. 8444.
[2] "Zoologist," 1863, p. 8841.
[3] Rodd, "List of the Birds of Cornwall," 2nd ed. p. 11.

common in the Crimea and in Hungary, extending eastward to Western and North-western India, where it is plentiful,[1] and is found accidentally in Italy, Switzerland, and France. Mr. Howard Saunders has reason to believe that it has been met with in Southern Spain in winter, but Col. Irby is somewhat sceptical on the point.[2]

In Sir Oswald Mosely's "Natural History of Tutbury" (p. 385), it is reported that a pair of the North American Red-eyed Flycatcher (*Muscicapa olivacea*) appeared at Chellaston, near Derby, in May, 1859, and one of them was shot. If there was no mistake in the identification of the species, one can only suppose that the birds must have been brought over to this country in a cage, and contrived to effect their escape.

---

[1] *Cf.* Hume, "Journ. Asiatic Soc. Bengal," 1870, p. 116, and Blanford, "Ibis," 1870, p. 534.

[2] See his "Ornithology of the Straits of Gibraltar," p. 224.

THE SWALLOW.

(Hirundo rustica.)

FEW birds have attracted more attention in all countries and in all ages than the Swallows; and the habits of those species which annually visit the British Islands have been so thoroughly investigated and so frequently described, that little originality can be claimed for the remarks which I have now to offer.

There are two points, however, in the natural history of these birds which do not appear to have received from their biographers so much attention as they deserve, viz., the nature of their food, and their geographical distribution. I have repeatedly been asked, "What do Swallows feed upon?" and "Where do Swallows go in winter?" To these two questions I will now endeavour to reply, believing that an exposition of such facts as have been ascertained on these points will be more acceptable to the reader than a repetition of what has been so frequently published on the subject of habits, haunts, dates of arrival, and other minor details.

First, then, as regards food. Dr. Jenner found that Swallows on their arrival in this country, and for some time afterwards, feed principally on gnats; but that their favourite food, as well as that of the Swift and Martin, is a small beetle of the Scarabæus kind, which he found, on dissection, in far greater abundance in their stomachs than any other insect. A writer in the "Magazine of Natural History,"[1] Mr. Main, states that they take two species of gnat, *Culex pipiens* and *C. bifurcatus*; and Sir

---

[1] "Mag. Nat. Hist." vol. iv. p. 413.

Humphrey Davy saw a single Swallow capture four Mayflies that were descending to the water, in less than a quarter of a minute. Mr. Thompson says[1] that a correspondent of his, Mr. Poole, has found the mouths of young birds filled with *Tipulæ*, and that Mr. Sinclair, an accurate ornithologist, remarked a number of Swallows flying for some time about two pollard willows, and on going to the place ascertained that the object of pursuit was hive bees, which, being especially abundant beneath the branches, he saw captured by the birds as they flew within a few yards of his head. The assertion that Swallows take honey bees was long ago made by Virgil, and, though not often noticed by writers on British Birds, the fact has several times been corroborated. A writer in the "Field Naturalist's Magazine" for 1834 (p. 125), stated that, having observed some Swallows seize bees in passing his hives, he shot them, and on opening them carefully, found that,

---

[1] "Nat. Hist. Ireland" (Birds), i. p. 377.

although they were literally crammed with drones, there was not a vestige of a working bee. We learn from Wilson[1] that in the United States bees constitute part of the ordinary food of the Purple Martin; and the Sand Martin has been observed to prey upon the common wasp. Gilbert White remarked that both Swifts and Swallows feed much on little Coleoptera, as well as on gnats and flies, and that the latter birds often settle on the ground for gravel to grind and digest their food. At certain times in the summer he had observed that Swifts were hawking very low for hours together over pools and streams, and, after some trouble, he ascertained that they were taking *Phryganeæ*, *Ephemeræ*, and *Libellulæ* (Cadew-flies, May-flies, and Dragon-flies), that were just emerged out of their aurelia state. The indigestible portions of their food are rejected in the shape of small pellets, just as with the birds of prey. *Apropos* of these observations, Mr. J. H. Gurney, in

---

[1] "American Ornithology."

October, 1871, wrote me as follows:—"The perusal of your interesting remarks relative to the food of the Chimney Swallow, and especially with reference to its bee-eating propensities, induces me to send you a note of an analogous habit of which I have heard, in one instance, in the Common Swift. An intelligent shepherd in Norfolk, with whom I am acquainted, and who keeps bees, states that a pair of Swifts which nested in the roof of his cottage were so destructive to his bees, by catching them on the wing when they happened to fly rather higher than usual, that he at length destroyed the Swifts in order to free his bees from their attacks. With reference to the food of the House Martin, I may mention that some years since, as I was watching some of these birds skimming over a roadside pond early in the month of May, one of them, as it flew past me, dropped at my feet a water beetle of the genus *Dytiscus*, nearly, if not quite, half an inch in length. Possibly it had captured a prey too large to be conveniently swallowed." All the

Hirundinidæ drink upon the wing, and are perhaps the only birds that do not alight for this purpose, unless perhaps the Terns and some of the Gulls may be also exceptions to the general rule.

With regard to their winter quarters and geographical distribution, it will be best to trace the movements of each species separately.

The Chimney Swallow (*Hirundo rustica*), whose early appearance in the spring is only preceded by that of the Sand Martin, spends at least six months of the year with us, and in some years more than seven months. The period of its visit, however, may be said briefly to extend from April to October. Between these two months the bird is found generally distributed throughout Europe, going as far north as Iceland[1] and Nova Zembla,[2] and penetrating even into Siberia and Amurland.[3]

---

[1] Professor Newton's Appendix to Baring Gould's "Iceland," p. 408.
[2] Gillet, "Ibis," 1870, p. 306.
[3] Von Schrenck, "Reise in Amurland."

The only Swallow hitherto observed in Greenland—and that only on two occasions—is, according to Professor Reinhardt, the American Swallow, *Hirundo rufa* of Bonaparte. Now, Bonaparte identifies this (Geogr. and Comp. List, p. 9) with *H. rufa* of Gmelin, and Professor Baird considers Gmelin's bird to be the South American species, for which *H. erythrogaster* of Boddaert is the oldest name. If this identification be correct, one would certainly expect the bird found in Greenland to be the North American species, *H. rufa* of Vieillot, not Bonaparte, now generally better known by its older name, *H. horreorum* of Barton. The late Mr. Wheelwright observed the Common Swallow in Lapland, where he saw it hawking about over the high fells at Quickjock, and he fancied it was even commoner there than at Wermland, in Sweden, where it is also an annual summer visitant.[1] Throughout Europe generally, as already remarked, it is everywhere distributed in

---

[1] "A Spring and Summer in Lapland," p. 281.

summer, and in the countries bordering the Mediterranean it is especially abundant at the periods of migration in spring and autumn. Mr. Wright has observed it arriving in Malta in great numbers from the south early in March, and again, on its return southwards in autumn, it is common over the island until October. On the island of Filfla, a few miles south of Malta, the same observer has noticed it in May. At Gibraltar and in Spain Mr. Howard Saunders has detected it as early as February, making its way north; and, as an instance of how these delicate birds at times get blown out of their course by adverse winds, it may be remarked that Prince Charles Bonaparte saw Swallows and Martins at sea 500 miles from Portugal and 400 miles off the coast of Africa. Sir William Jardine has recorded the presence of the Swallow at Madeira, and Mr. Osbert Salvin, writing on May 28 ("Ibis," 1859, p. 334), says: "Some Swallows came on board when we were 180 miles north-west of the Azores, so that it is probable that the bird is found in these islands."

On the Senegal River and at Sierra Leone it may be seen all the year round, but is less numerous there from June to September.[1] On the West Coast of Africa the Swallow appears to travel as far south as the island of St. Thomas on the equator, where Mr. Yarrell states it has been met with in January and February.

In Tangier and Eastern Morocco Mr. Tyrwhitt Drake says the Swallow is found throughout the year, but the Martin and Sand Martin, he believes, do not winter there. ("Ibis," 1862, p. 425.)

The Swallow has been noticed as plentiful at Tripoli in the middle of March (Chambers, "Ibis," 1867, p. 99), and Mr. Osbert Salvin observed it in Algeria, between Constantine and Batna, where he found several nests among the rafters of an open shed. According to the Rev. Canon Tristram—than whom there is no better authority on the subject of North African and Palestine birds—a few pairs of Swallows remain all the winter in each oasis in North Africa, wher-

---

[1] Tudbury, "Mag. Nat. Hist." vol. v. p. 449.

ever there is water or marsh; but none of those which he observed were in mature plumage, and it is therefore presumed that only the younger and weaker birds stay behind. The Arabs informed him that for one Swallow they have in winter they have twenty in summer, and that they usually retire about the end of November, returning in February. In November, also, they have been observed to be common at Alexandria and Cairo (E. C. Taylor, "Ibis," 1859, p. 47); and on the 5th of November, when leaving Aden, Mr. Swinhoe remarked that a few Swallows followed the ship, apparently bound for the Indian coast. According to the observations of Mr. E. C. Taylor ("Ibis," 1867, p. 57), this species reappears in Egypt about March 25, and is common at Cairo and Damietta in April. Rüppell, in his " Systematische Uebersicht der Vögel Nordost-Afrika's," includes the Swallow (p. 22) as being found in Egypt, Nubia, and Abyssinia; and as regards the last-named country, Mr. Blanford has remarked[1]

---

[1] " Geology and Zoology of Abyssinia," p. 347.

that it is common everywhere, and that he found it especially abundant on the shores of Annesley Bay in June.

Continuing a search for this species southward along the East Coast of Africa, it will be found that, according to the observations of Mr. Ayres in Natal, the Swallow arrives in that colony in great numbers in November, congregating and leaving again in March and April. Mr. Layard found it to be an annual winter visitant to the Cape Colony, and on one occasion when sailing from New Zealand to the Cape of Good Hope, on the 28th of November, he saw a Swallow and a Sand Martin fly about the ship for some time. He was then in lat. 33° 20′, long. 31° 50′, and about 290 miles from the Cape. Several insects (*Libellula, Agrostis*, and *Geometra*) were caught on deck, and we may presume, therefore, that the birds found sufficient food to support them at that distance from land.

Passing eastward through Sinai and Palestine, where Canon Tristram has observed the Swallow in December,[1] we learn from Mr. Blyth that it is common in the north-west provinces of India during the winter months. Capt. Beavan saw it at Darjeeling in 1862, and Maunbhoom in 1864-65, where both old and young were very common in January and February, hawking over rice-kates and near tanks. In Northern Japan it was observed by Capt. Blakiston, and in North China by Mr. Swinhoe. Referring to a species of Swallow which he observed in Formosa, Mr. Swinhoe says ("Ibis," 1863, p. 255): "In its habits, in nest and colour of eggs, &c., this bird entirely agrees with the European *H. rustica;* yet in size it is always smaller, and in minor personal features different. It ranges in summer from Canton to Pekin, and Mr. Blyth assures me that it is identical with specimens procured in winter in Calcutta; hence I infer that the birds which visit China in spring, and uniformly leave again in autumn, return to hybernate in the warm plains of India."

---

[1] "The Land of Israel," p. 105.

Mr. Blyth has remarked ("Ibis," 1866, p. 336), "that the average of adult Swallows from the Indian region and China are smaller than the average of European examples, to the extent sometimes of an inch in length of wing; but some Indian are undistinguishable from European specimens."

Dr. Jerdon, in his "Birds of India," says: "On carefully comparing specimens from England and Algiers in the museum at Calcutta with Indian specimens from various parts of the country, I can detect no difference."

In a notice of the birds of the Andaman Islands which appeared in the "Ibis" some years since, Capt. Beavan remarked that the European Chimney Swallow visits these islands at certain seasons, and is not at all uncommon.

There is no evidence that it ever visits Australia; but Mr. Gould has described a Swallow from Torres Straits under the name *Hirundo fretensis*, which is certainly very like our well-known *H. rustica*, and might be a young bird of that species in autumn plumage. It is singular that no Swallows visit New Zealand. It cannot be that the islands are too distant from Australia, where several species of Swallow abound, because, as Mr. Layard has remarked, two, if not three, species of Cuckoo (*Eudynamys taitensis* and *Chrysococcyx lucidus*) perform the journey in their annual migration twice a year.

The attachment of Swallows to the neighbourhood of water at roosting-time—which formerly led to the supposition that they actually retired under water for the winter—may be easily accounted for by the circumstance that the willow branches not only afford them most convenient perches, but enable the birds to crowd close together, and so secure greater warmth to individuals than they could possibly enjoy if each roosted upon a separate twig in trees or shrubs of different growth.

THE MARTIN.

(*Hirundo urbica.*)

ALTHOUGH arriving in this country somewhat later than the Swallow, the Martin may be said to have nearly the same geographical range. Mr. Yarrell thought that the Swallow did not go so far north as the Martin,[1] but both are found in summer in Iceland and the Faroe Isles. Mr. Dann remarked that there was no want of food for them in Norway and Lapland, as the morasses in the sheltered valleys swarm with insects. During the season that it is absent from England it resides in North Africa, Egypt,

---

[1] " History of British Birds," vol. ii. p. 251 (3rd ed.)

Nubia and Abyssinia, Palestine, Arabia, and North-west India. Capt. Irby states ("Ibis," 1861, p. 233) that it is common in the cold season in Oudh, and Col. Tickell observed great numbers at Moulmein ; but they appeared from time to time, and not constantly, like *H. rustica*.[1] With regard to Palestine, it seems probable that the Martin spends the greater portion of the year there, for Canon Tristram found it breeding in colonies on the sheltered faces of cliffs in the valleys of Northern Galilee. Mr. Wright says ("Ibis," 1864, p. 57) that in Malta it is seen at the same seasons as the Swallow, but stays part of the winter, when *H. rustica* has departed. Dr. Giglioli observes that it arrives at Pisa at the end of March, at which time it has also been noticed at Gibraltar.

The movements of this bird and others of the genus have been concisely illustrated by Mr. Forster in a communication to the Linnæan Society, in the following table, giving the mean date of arrival:

---

[1] Journ. As. Soc. Beng. xxiv. p. 277.

	Naples.	Rome.	Pisa.	Vienna.	Bruges.	London.
Swallow . . .	Feb. 27.	Mar. 3.	Mar. 5.	Mar. 25.	Ap. 5.	Ap. 15.
Sand Martin .	Ap. 3.	Ap. 5.	Ap. 8.	Ap. 12.	Ap. 25.	Ap. 25.
House Martin .	Ap. 10.	Ap. 15.	Ap. 16.	Ap. 20.	May 1.	May 1.
Swift	Ap. 15.	Ap. 18.	Ap. 20.	Ap. 23.	Ap. 30.	May 3.

The spring tide of migration appears to set in along the entire coast-line of the Mediterranean, and in a direction almost due north. I do not remember to have seen any record of the occurrence of the Martin on the west coast of Africa, although there seems to be no reason why it should not accompany the Swallow there in winter.

Both species will rear two broods in a season; and this fact, doubtless, will account for the prolonged stay in autumn of the later fledged birds, which are not sufficiently strong on the wing to join the main body of emigrants at the usual time of their departure.

THE SAND MARTIN.

(Cotyle riparia.)

THIS little bird has a much more extensive range than either of the foregoing species, being found in the New as well as in the Old World. In British North America M. Bourgeau obtained both birds and eggs on the Saskatchewan plains. Dr. Coues met with it in Arizona, and Professor Baird has recorded it from California. He says: " It furnishes almost a solitary instance amongst land birds of the same species inhabiting both continents permanently, and not as an accidental or occasional visitor in either."[1] Mr. H. E. Dresser found it common in Southern Texas, and Mr. O. Salvin obtained several specimens in Guatemala. It has even been met with in the Bermudas, 600 miles from Cape Hatteras, the nearest point of the North American coast.[2] .

In Europe the Sand Martin generally makes its appearance in the spring somewhat earlier than any of the other Swallows, and departs sooner. From different stations on the Mediterranean large flocks have been observed at the period of the vernal migration winging their way northward, returning even in greater numbers in the autumn. Mr. O. Salvin saw this species between Tunis and Kef during the third week in March. Canon Tristram, who found it abundant in Palestine in the sandy banks of the Jordan, has suggested that it is double-brooded, since he found it nesting in Egypt in February. The same observer met with it in November on its autumn migration through the Sahara. When

---

[1] "Birds of North America," p. 313.
[2] Jones's "Naturalist in Bermudas," p. 34.

passing down the Red Sea, early in November, Mr. Swinhoe saw numerous Sand Martins, which followed the ship for some days, and on arriving at his destination found these birds very common about the marshes at Takoo and before Tientsin in North China. Dr. Leith Adams says[1] that Sand Martins build in numbers along the banks of the Indus, and that in consequence in some places the banks are quite riddled with their holes. Hence it will be seen that this delicate little bird enjoys a more extensive range than any other species of the family.

Before leaving this country in autumn, they assemble in vast flocks, and go through a variety of evolutions on the wing, as if practising for a long flight, alighting from time to time upon the ground, or on willows or reeds by the river-side, to rest. Swallows and Martins do the same, but never congregate—so far as I have observed—in such large numbers.[2]

---

[1] "Wanderings of a Naturalist in India," p. 49.
[2] See "The Birds of Middlesex," p. 126.

The Purple Martin (*H. purpurea*) of America is recorded to have been procured once at Kingstown near Dublin; and Yarrell included it in his "History of British Birds," relying on a statement that two specimens had been shot at Kingsbury Reservoir, in Middlesex, in September, 1842. It has since been ascertained, however, that he was misinformed on the subject. A specimen of this bird, said to have been shot near Macclesfield, was sold at Stevens's, with other birds from the Macclesfield Museum, on the 14th June, 1861, and realized twenty-eight shillings. With these exceptions, so far as I am aware, no other instance of its occurrence in Europe has been published.

THE COMMON SWIFT.

(*Cypselus apus.*)

TO ordinary observers a Swift appears so much like a Swallow, that the only difference discernible by them is a difference of colour. To the inquiring naturalist, however, a much more important distinction presents itself in the peculiar and remarkable anatomy of the former bird. Not only has it a greater extent of wing, moved by larger and more powerful muscles, but the structure of the foot is curiously adapted for climbing within the narrow crevices which are usually selected as nesting-places. In the Swallow and other *Hirundines* the toes are long and slender—three in front and one behind in the same plane, as is usual with insessorial birds. In the Swift we find the toes short and stout, and all four directed forwards; the least toe (which should be the hind one) consisting of a single bone, and the other three of only two bones apiece—a peculiar construction, but well adapted for the purposes for which the feet are employed.

This singularity of structure has induced naturalists to consider the Swifts (for there are several species) generically distinct from the Swallows; and the former, therefore, are now placed by common consent in the genus *Cypselus*, a name adopted from Aristotle, and suggested by Illiger, as indicating the bird's habit of hiding its nest in a hole.

The remarks which have been made upon food in the case of the Swallows, apply equally in the case of the Swifts. The latter have so frequently been observed in localities presenting very different species of insects, and sweeping in the summer evenings through the midst of little congregated parties of various kinds, that there is little doubt that the nature of the food differs very considerably. In corroboration of this it has been shown that anglers have repeatedly captured these birds with artificial trout-flies of very different appearance.[1] Isaak Walton informs us that Swifts were in his time taken in Italy with rod and line; and, according to Washington Irving, one of the sports of the Alhambra was angling for swallows from its lofty towers.[2] There are several species of Swifts distributed throughout the world, but only two visit the British Islands, and of these one is but a rare and accidental visitant.

The Common Swift is the last of the *Hirundines* to arrive in this country, and the first to leave it. Its habits are very different from those of the Swallows. As a rule it makes no nest, but only lines a hole, into which it creeps; it lays but two eggs (rarely three), instead of five or six like the Swallows; it rears but one brood in the summer, instead of two, or even three, as Swallows often do. The late Mr. J.

---

[1] Thompson, "Nat. Hist. Ireland" (Birds), i. p. 377.
[2] Irving, "Tales of the Alhambra."

D. Salmon described[1] some nests of the Swift which he found at Stoke Ferry, Norfolk, and which were composed of bits of straw and dry grass, " closely interwoven and held firmly together by an adhesive substance very much resembling glue, and so disposed round the inner edge of the nest as to hold the straws in their places; the whole forming quite a cup of an oval shape, of about four inches in length, not very deep." I have often observed the straw and dry grass, with the addition of feathers, but never noticed the "adhesive substance." Gilbert White thought that the Swift paired on the wing. They may do so occasionally; but, from what I have observed, I feel sure that they pair much oftener in the hole which has been selected to nest in.

Although usually preferring lofty towers and church turrets, the Swift frequently nests under eaves at a comparatively short distance from the ground; and I have had excellent opportuni-

---

[1] "Mag. Nat. Hist." 1834, vol. vii. p. 462.

ties for some years past of observing Swifts during the breeding season under the eaves of some old cottages in Sussex and Middlesex. By means of a short ladder I have been enabled to inspect many nests both before and after the young were hatched; and, out of a score or more examined, seldom more than one contained three eggs. Sometimes I observed that Sparrows were ejected and their nests appropriated, amidst much remonstrance and screaming; but, as a rule, I have found that Swifts, having once reared their young safely in a new locality, will return to the same hole year after year. Birds have been marked by having their claws cut, and, on being set at liberty, were caught the following year in the holes from which they had first been taken. Unlike most insectivorous birds, which bring but a single insect (or at most two or three) to the nest at a time, the Swift visits its young less frequently in the day, but brings a large store at each visit. The mouth is often so crammed with small black flies, that the bird presents the appearance of having a pouch under the chin, from which it ejects the insects in a lump the size of a boy's marble.

As a general rule, the Swift is not observed in this country before the third week in May, and is seldom seen after the third week in August. It is found throughout the mainland of the British Islands, and breeds also in Mull and Iona, but not in Orkney or Shetland, nor in the Outer Hebrides. It does not travel quite so far north as either the Chimney Swallow or the Martin, but the late Mr. Wolley saw it on the Faroes,[1] and Mr. Wheelwright frequently observed it hawking over the high fells at Quickjock, Lapland, during the summer.[2]

If we look for the bird during the months that it is absent from Great Britain, we find that it is very abundant at the Cape of Good Hope in winter, arriving about September 5, and departing northwards in April. It is seen in

---

[1] "Contributions to Ornithology," 1850, p. 109. It is not included by Herr Müller in his "Bird Fauna of the Faroes."
[2] "A Spring and Summer in Lapland," p. 281.

Natal more or less all the year round, but more plentifully during the summer.[1] The climate of Lower Egypt is apparently too cold for this species in winter, but at that season it is resident and abundant in Upper Egypt.[2]

As winter disappears it gradually moves northward, and a month before it arrives in England it is found in some numbers along the entire coast-line of the Mediterranean. Mr. Osbert Salvin saw it at Tunis on the 8th March, and subsequently numerous at Algiers. In the middle of March, Mr. Chambers found it plentiful at Tripoli, and at the end of the same month it was observed by Mr. Howard Saunders at Gibraltar. In the middle of April, Lord Lilford remarked that it was common in the neighbourhood of Madrid; about which time, according to Messrs. Elwes and Buckley ("Ibis," 1870, p. 200), it usually makes its appearance in Turkey, arriving there doubtless from the Ionian Islands,[3]

---

[1] Ayres, "Ibis," 1863, p. 321.
[2] E. C. Taylor, "Ibis," 1867, p. 56.
[3] Lord Lilford, "Ibis," 1860, p. 234.

Egypt and Palestine, where it is said to appear in the last week of March.[1]

From Spain, through France, to England is but a short journey for a bird with powers of wing like the Swift; and hence one is not surprised to see hawking over the South Downs in May the birds which but a week previously were circling round the Moorish towers of Spain. Its return southward in autumn is apparently by the same route as that chosen for its northward journey in spring, and in this respect it differs in habit from many other species.[2]

In India its place is to a certain extent taken by a non-migratory species, *Cypselus affinis*, but it has nevertheless been met with in that country. An Indian specimen was received from Dr. Jerdon, presumably from the north-west.[3] It has also been forwarded from Afghanistan,[4] and

---

[1] Tristram, "Ibis," 1865, p. 77.

[2] In the Grey Phalarope we have a notable instance of a contrary habit. This bird passes through England on its way southward in autumn, but invariably selects some other route on its return northward in spring.

[3] Blyth, "Ibis," 1866, p. 339.

[4] Blyth, "Ibis," 1865, p. 45.

Dr. Stoliczka found it at Leh, in Western Thibet. I am aware that some naturalists have expressed doubts as to the identity of the Swift found at the Cape of Good Hope with *Cypselus apus;* but, after an examination of several examples of the African bird, I have been unable to discover that it differs in any material respect from our well-known summer migrant.

THE ALPINE SWIFT.

(*Cypselus alpinus.*)

SO rare a visitant to this country is the Alpine Swift that not more than a score of individuals have been met with since the first specimen was captured in 1820. In that year a bird of this species was killed at Kingsgate, in the Isle of Thanet, during the month of June, and since that time the following examples are recorded to have been met with:—

One, Dover, Aug. 20, 1830; "Note-book of a Naturalist," p. 226.

One, Buckenham, Norfolk, Oct. 13, 1831; Yarrell, "Hist. Brit. Birds," vol. ii. p. 266.

One, Rathfarnham, near Dublin, March, 1833; "Dublin Penny Journal," March, 1833. Yarrell, "Hist. Brit. Birds," vol. ii. p. 266.

One, Saffron Walden, Essex, July, 1838; Macgillivray, "Hist. Brit. Birds," iii. p. 613.

One, Leicester, Sept. 23, 1839; Macgillivray, "Hist. Brit. Birds," iii. p. 613.

One, seen forty miles west of Land's End, in June, 1842; Couch, "Cornish Fauna," p. 147.

One, Cambridge, May, 1844; E. B. Fitton, "Zoologist," 1845, p. 1191.

One, near Doneraile, co. Cork, June, 1844; Thompson, "Nat. Hist. Ireland" (Birds), vol. i. p. 418.

One, St. Leonard's-on-Sea, Oct., 1851; Ellman, "Zoologist," 1852, p. 3330.

One, Mylor, Cornwall, 1859; Bullimore, "Cornish Fauna," p. 24.

One, Hulme, near Manchester, Oct. 18, 1863; Carter, "Zoologist," 1863, p. 8846.

One seen at Kingsbury Reservoir, Aug. 1841,

and one shot near Reading the next day; Harting, " Birds of Middlesex," p. 128.

One, near Lough Neagh, May, 1866; Howard Saunders, " Zoologist," 1866, p. 389.

One, near Weston-super-Mare ; Cecil Smith, " Birds of Somersetshire," p. 287.

Several seen, Isle of Arran, July, 1866 ; H. Blake Knox, " Zoologist," 1866, p. 456.

Several seen, Achill Island ; H. Blake Knox, " Zoologist," 1866, p. 523.

One, near the Lizard, Cornwall; Rodd, " List of the Birds of Cornwall," 2nd ed. p. 23.

One, Aldeburgh, Suffolk, Sept. 8, 1870; Hele, " The Field," Sept. 17, 1870.

One seen, Colchester, June 8, 1871; Dr. Bree, " The Field," June 17, 1871.

One seen, South Point, Durham, July 24, 1871 ; G. E. Crawhall, " The Field," Aug. 5, 1871.

In all the above instances the birds were shot, except where stated to have been seen only.

The term "Alpine Swift" is unfortunately a misnomer, since the bird is in no way confined to the immediate neighbourhood of the Alps. The name "White-bellied Swift" is not inappropriate, as indicating a peculiarity which distinguishes it from the common species. It is a migratory bird, like the last-named, and, like it, visits the Cape of Good Hope in winter, and penetrates into North-west India.

It is a summer migrant in Palestine, where Canon Tristram observed it nesting near Mar Saba, and in the tremendous ravine above the site of Jericho. It arrives at Constantinople from its winter quarters towards the end of April, and is common in Corfu from May to September, nesting annually in the Citadel Rock (Lord Lilford, " Ibis," 1860, p. 234). It breeds in great numbers along the Etruscan coast, and is occasionally seen at Pisa (Dr. Giglioli, "Ibis," 1865, pp. 51-52). It has been observed on passage in Tangier and Eastern Morocco, and Mr. O. Salvin remarked that it was common about the plains of the Salt Lake district, Eastern Atlas, and breeding in most of the rocks of that country ("Ibis," 1859, p. 302).

Mr. Howard Saunders saw hundreds at Gibraltar towards the end of March, and in June it was observed by Lord Lilford amongst the peaks of the Sierra near San Ildefonso. To England, as we have said, it rarely strays. In habits it is described, by those who have had opportunities for observing it, as resembling very much the Common Swift. Like this species, it nests in holes and crevices, and lays two white eggs of a similar shape to those of its congener, but much larger. Its cry is said to be very different. Its vastly superior size and white belly serve at all times to distinguish it from the smaller and more sable bird with which we are so familiar.

The Spine-tailed Swift (*Acanthylis caudacutus*), a bird which is found in Siberia, Persia, India, China, and Australia, has in one single instance been met with in the British Islands. A specimen was killed at Great Horkesley, near Colchester, on July 8, 1846, as recorded in the "Zoologist" for that year (p. 1492), and was fortunately examined in the flesh by Messrs. Yarrell, Fisher, Hall, Doubleday, and Newman.

THE NIGHTJAR.

(Caprimulgus europæus.)

IN order of date, the Nightjar is one of the latest of the summer birds to arrive, being seldom seen before the beginning of May, although, as in the case of other species, one now and then hears of an exceptionally early arrival. In 1872, for example, Mr. Gatcombe informed me that he had seen a Nightjar in the neighbourhood of Plymouth on the 10th of April, at least a month earlier than the usual time of its appearance. By the end of September, or the first week in October, these birds have returned to their winter quarters in North Africa. Colonel Irby, in his recently-published volume on the "Ornithology of the Straits of Gibraltar," states, on the authority of M. Favier, that Nightjars cross the Straits from Tangiers to Gibraltar in May and June, and return the same way between September and November. They have been seen on the passage. Dr. Drummond informed the late Mr. Thompson of Belfast,[1] that when H.M.S. "San Juan," of which he was surgeon, was anchored near Gibraltar, in the spring of the year, a few Nightjars flew on board. During the passage of H.M.S. "Beacon" from Malta to the Morea, in the month of April, some of these birds appeared on the 27th, and alighted on the rigging. The vessel was then about fifty miles from Zante (the nearest land), and sixty west of the Morea.

They came singly, with one exception, when two appeared in company. A couple of them were shot in the afternoon. A few others had

---

[1] See Thompson's "Nat. Hist. Ireland" (Birds), vol. i. p. 423.

been seen about the vessel on the two or three days preceding. On the evening of the 1st of June, two were killed and others seen in the once celebrated but now barren and uninhabited island of Delos.

The Nightjar, although tolerably dispersed throughout North Africa during certain months of the year, does not, apparently, travel so far down the east or west coasts as many of our summer migrants do. In Egypt and Nubia, according to Captain Shelley,[1] it is only met with as a bird of passage, but how much further south it goes he does not say. Mr. Blanford did not meet with it in Abyssinia, where its place seems to be taken by two or three allied species.[2] The same remark applies as we proceed eastward. In Syria and Palestine, Canon Tristram did not observe the European Nightjar, but found a smaller and lighter-coloured species, on

---

[1] "Birds of Egypt," p. 174.
[2] "Observations on the Geology and Zoology of Abyssinia," p. 336.

which he has bestowed the name *Caprimulgus tamaricis.*[1]

Between the months of April and October, our Nightjar is generally dispersed throughout the British Islands, even to the north of Caithness, extending also to the inner group of islands, but not reaching the Outer Hebrides. Mr. Robert Gray, of Glasgow, reports that it is not uncommon in Islay, Iona and Mull, and also in Skye, in all of which islands eggs have been found.

Stragglers have been observed in summer and autumn for several years in Shetland. The late Dr. Saxby saw it at Balta Sound about the end of July, skimming over the fields, and now and then alighting on the dykes, but he regarded its appearance in Shetland as merely accidental.

In Ireland this bird is considered to be a regular summer visitant to favourite localities in all quarters of the island, but of rare occurrence elsewhere.[2]

---

[1] "The Land of Israel," p. 250.   [2] Thompson, *op. cit.*

In colour this bird resembles a large moth, being most beautifully and delicately streaked and mottled with various shades of black, brown, grey, and buff, but in appearance it is not unlike a hawk, having long pointed wings more than seven inches in length, and a tail about five inches long. The male differs from the female in having a large heart-shaped spot upon the inner web of the first three quill feathers, and broad white tips to the two outer tail feathers on each side.

The mottled brown appearance of the bird when reposing either on the ground or on the limb of a large tree, is admirably adapted to screen it from observation even within a few yards of the observer. It delights in furzy commons, wild heathery tracts, and broken hilly ground covered with ferns, particularly in the neighbourhood of woods and thickets, and is especially partial to sandy soils. I have frequently seen this bird upon the bare sand, either in a sandpit or under the lee of a furze-bush, where it appeared to be basking in the sun, and from the disturbed appearance of the soil in some places, I imagine that it dusts itself as the Skylark does, to get rid of the small parasites with which, like many other birds, it is infested. On the 16th of May this year, at Uppark, Sussex, I found one asleep on the carriage drive within twenty yards of the house. The gravel was quite warm, and the bird was so loth to be disturbed that I almost succeeded in covering it with my hat before it took wing. On another occasion in September, when strolling along the beach near Selsea, I came suddenly upon a Nightjar sitting below high-water mark on the warm shingle, where it appeared to be thoroughly enjoying the afternoon sun. It dozes away the greater part of the day, and if disturbed only flies a short distance before re-alighting. Its loud and peculiar whirring note, reminding one of the noise made by a knife-grinder's wheel, is never heard until the evening, when, in districts where the bird is common, it resounds far and near.

There is something occasionally quite ventriloquial in the sound, caused by the bird turning its head from side to side, both up and down, and scattering, as it were, the notes on every side.

It makes no nest, but scraping a hollow on the bare ground deposits two ellipse-shaped eggs beautifully mottled with two shades of grey and brown, and quite unlike those of any other British bird. The young are hatched in about a fortnight or rather more, and until fully fledged their appearance is singularly ugly. They are covered with a grey down, and their enormous mouths and large prominent eyes give them an expression which is almost repulsive. By pegging the young down with long "jesses," as one would a Hawk, I have secured them until fully fledged, the old birds feeding them regularly; but on taking them home and turning them into an aviary I could not succeed in keeping them long alive, owing to the difficulty in procuring suitable food, and my inability to give them constant attention.

During the month of September, when shooting amongst low underwood and felled timber,

I have not unfrequently disturbed a Nightjar, and on such occasions, when flying away startled, its flight so much resembles that of a Hawk that I have twice seen a keeper shoot one, exclaiming, " There goes a Hawk ! " I was not a little surprised one day at finding one of these birds in the middle of a turnip-field. We had marked down some birds at the far end, and the dogs were drawing cautiously on when one of them flushed a Nightjar, which my friend immediately shot—in mistake, as he afterwards said, for a Woodcock.

Notwithstanding what has been said to the contrary, the Nightjar, Night-hawk, Fern Owl, or Goatsucker, as it is variously called in different parts of the country, is one of the most inoffensive birds imaginable. By farmers it is accused of robbing cows and goats of their milk, and by keepers it is remorselessly shot as "vermin;" but by both classes its character is much maligned. Its food is purely insectivorous, and it is as incapable of sucking milk as it is of carrying off and preying upon young game birds. The mistake in the former case must have arisen in this way. The habits of the bird are crepuscular. It is seldom seen in broad daylight unless disturbed, but as soon as twilight supervenes and moths and dor-beetles begin to be upon the wing, it comes forth from its noonday retreat and is exceedingly busy and active in the pursuit of these and other insects. Montagu says he has observed as many as eight or ten on the wing together in the dusk of the evening, skimming over the surface of the ground in all directions, like Swallows in pursuit of insects. Cattle, as they graze in the evening, disturb numerous moths and flies, and the Nightjar, unalarmed by the animals, to whose presence it becomes accustomed, dashes boldly down to seize a moth which is hovering round their feet, or a fly which has settled upon the udder. Being detected in this act in the twilight by unobservant persons, the story has gone forth that the Goatsucker steals the milk.

From the keeper's point of view it is a Night-

hawk in the worst sense of the word, a hawk that under cover of the night flits noiselessly but rapidly by and carries off the unsuspecting chick. But here again the observer has been misled by appearances, associating the pointed wings and long tail with the idea of a hawk, entirely overlooking the small slender claws and mandibles, which are quite unequal to the task of holding and cutting up live and resisting feathered prey, and entirely also overlooking the fact that at the time the Nightjar is abroad, the young pheasants and partridges are safely brooded under their respective mothers.

Attentive observation of its habits, and examination of numerous specimens after death, have revealed the real nature of its food, which consists of moths, especially *Hepialus humuli*,[1] which from its white colour is readily seen by the bird, fernchafers and dor-beetles. Macgil-

---

[1] Mr. Robert Gray of Glasgow has seen it in grass fields, cleverly picking ghost-moths (*Hepialus humuli*) off the stems, from the points of which these sluggish insects were temptingly hanging. But as a rule, he adds, the Nightjar captures its prey while in flight.

livray says: "The substances which I have found in its stomach were remains of coleopterous insects of many species, some of them very large, as *Geotrupes stercorarius*, moths of great size also, and occasionally larvæ. I have seen the inner surface slightly bristled with the hairs of caterpillars, as in the Cuckoo." He adds, "as no fragments of the hard parts of these insects ever occur in the intestine, it follows that the refuse is ejected by the mouth." From its habit of capturing dor-beetles, the bird in some parts of the country is known as the Dor-hawk. Wordsworth has referred to it by this name in the lines—

> "The busy Dor-hawk chases the white moth
> With burring note."

Elsewhere it is called the Eve-jar, and Churn-owl. The latter name is bestowed by Gilbert White in his "Naturalist's Summer Evening Walk":—

> "While o'er the cliff the awaken'd Churn-owl hung,
> Through the still gloom protracts his chattering song."

In his 37th Letter to Pennant, the same author

refers to it as "the *Caprimulgus*, or Fern-owl," and gives an agreeable account of its movements as observed by himself.

Amongst other things he says :—" But the circumstance that pleased me most was, that I saw it distinctly more than once put out its short leg while on the wing, and by a bend of the head deliver somewhat into its mouth. If it takes any part of its prey with its foot, as I have now the greatest reason to suppose, it does these chafers. I no longer wonder at the use of its middle toe, which is curiously furnished with a serrated claw."

Yarrell has figured the foot, in a vignette to his work on British Birds, in order to show this peculiarity of structure, the use of which has puzzled so many.

The correctness of the view expressed by Gilbert White and confirmed by other authors,[1] has been disputed on the ground that many other birds, as Herons, Gannets, and, I may

---

[1] See Atkinson's "Compendium of Ornithology," p. 108, and Stanley's "Familiar History of Birds," p. 260.

add, Coursers, have a pectinated claw upon the middle toe, and yet do not take insects upon the wing, or even seize their prey with their feet.

It has been ingeniously suggested that perhaps the serrated claw may be used for brushing away the broken wings and other fragments of struggling insects which doubtless adhere occasionally to the basirostral bristles with which the mouth of this bird is furnished. This is very possible; at the same time it may be observed that Hawks, Parrots, and other birds habitually cleanse the bill and sides of the gape with their feet, and yet have no pectination of the middle claw.

A theory advanced by Mr. Sterland,[1] and endorsed by Mr. Robert Gray,[2] is that since the Nightjar sits *lengthwise* and not *crosswise* upon a bough, the serrated claw gives a secure foothold, which in so unusual a position could not be obtained by grasping. But to this theory the objection above made also applies, namely, that many birds, such as Coursers and Thick-knees,

---

[1] " The Birds of Sherwood Forest," p. 172.
[2] " The Birds of the West of Scotland," p. 212.

have serrated middle claws and yet are never seen to perch.

Some naturalists, and amongst others Bishop Stanley, have surmised that by means of its peculiarly-formed toes, the Nightjar is enabled to carry off its eggs, if disturbed, and place them in a securer spot, but should any such necessity arise, one would think that its large and capacious mouth, as in the case of the Cuckoo, would form the best and safest means of conveyance.

In the young Nightjar at first the peculiarity in question is not observable, and Macgillivray remarked that in a fully-fledged young bird shot early in September, the middle claw had only half the number of serrations which are usually discernible in the adult. He says :—
" All birds whose middle claw is serrated have that claw elongated, and furnished with a very thin edge. It therefore appears that the serration is produced by the splitting of the edge of the claw after the bird has used it, but whether in consequence of pressure caused by standing or grasping can only be conjectured." I have detected some confirmation of this in the case

of the common Thick-knee, or Stone Curlew, *Œdicnemus crepitans*, in some specimens of which I have remarked a very distinct serration of the middle claw, in others only the barest indication of it (the edge of the claw being very thin and elongated); in others again no trace of it.

The objections, however, which have been taken to the suggested use of the pectinated claw in the Nightjar, do not invalidate the statements which have been made by Gilbert White and other observers of the bird's movements and habits, for the homologous structure which is found to exist in certain species in no way related to each other, may well be designed for very different functions.

I do not find in the works of either Macgillivray or Yarrell any mention made of the peculiar viscous saliva which is secreted by this bird, and which reminds one of what is observable in the case of the Wryneck and the different species of Woodpecker. It no doubt answers the same purpose, namely, to secure more easily the struggling insects upon which its existence depends.

THE CUCKOO.

(Cuculus canorus.)

FROM numerous observations made by competent naturalists in different localities, it appears that the usual time of arrival of the Cuckoo in this country is between the 20th and 27th of April, and the average date of its appearance may be said to be the 23rd of that month, St. George's Day. In no instance, so far as I am aware, has the bird been heard or seen before the 6th of April. On that date in 1872 it was observed at Torquay, but this was considered by my informant an unusually early date at which to meet with it.

Between April and the end of August, it may be found generally distributed throughout the British Islands, even as far north as Orkney and Shetland. It is also a well-known visitor to the Outer Hebrides. On the European continent it occurs throughout Scandinavia and Russia, and is found in all the countries southward to the Mediterranean, which it crosses in the autumn for the purpose of wintering in North Africa. Eastward it extends through Turkey, Asia Minor, and Persia, to India, and according to Horsfield and Temminck, visits even Java and Japan.[1]

The Cuckoo does not pair, but is polygamous.

---

[1] The late Mr. Blyth thought that the Cuckoo found in Java by Dr. Horsfield was not the Common Cuckoo of Europe, but an allied race (*C. canoroides*, Müller, *optatus*, Gould), whose range extends eastward at least to China, and southward to Australia. If so, doubtless the same remark applies to Japan. *Cf.* "The Ibis," 1865, p. 31.

It is not unusual, soon after their arrival, to see a couple of male birds chasing a hen. The first eggs are seldom laid before the middle of May, or not until the birds have been here three weeks or a month. The egg, which is about equal in size to that of the Skylark, is very small, considering the bulk of the bird which lays it. It is white, closely freckled over with grey, or sometimes reddish brown, and generally has a few darker specks at the larger end. Instead of building a nest for itself, the Cuckoo deposits its eggs singly, and at intervals of a few days, in the nests of a variety of other birds, and leaves them to be hatched out, and the young reared, by the foster parents.

The nests in which the Cuckoo's eggs are most frequently deposited are those of the Hedge Sparrow, Meadow Pipit, Pied Wagtail, and Reed Warbler, but according to Dr. Thienemann, a great authority on the subject of European birds' eggs, they have also been found in the nests of the following very different species :—

Garden Warbler.
Blackcap.
Whitethroat.
Lesser Whitethroat.
Redstart.
Black Redstart.
Robin.
Reed Warbler.
Sedge Warbler.
Marsh Warbler.
Grasshopper Warbler.

Willow Wren.
Hedge Sparrow.
Common Wren.
Whinchat.
White Wagtail.
Grey-headed Wagtail.
Tawny Pipit.
Meadow Pipit.
Skylark.
Yellowhammer.

To this list Dr. Baldamus, from other sources, has added the following:[1]—

Red-backed Shrike.
Barred Warbler.
Nightingale.
Icterine Warbler.
Chiffchaff.
Great Reed Warbler.
Sedge Warbler.
Fire-crested Wren.

Tree Pipit.
Crested Lark.
Wood Lark.
Common Bunting.
Black-headed Bunting.
Greenfinch.
Linnet.
Russet Wheatear.

And lastly, in a foot-note to Mr. Dawson Rowley's article on the Cuckoo,[2] in which the above lists were quoted, Professor Newton has pointed out the authority which exists for inclu-

---

[1] "Naumannia," 1853, p. 307.

[2] On certain facts in the economy of the Cuckoo, "Ibis," 1865, pp. 178—186.

ding the following, at least occasionally, amongst the foster parents of the young Cuckoo :—

House Sparrow.	Mealy Redpoll.
Blue-throated Warbler.	Bullfinch.
Rock Pipit.	Jay.
Chaffinch.	Song Thrush.
Blackbird.	Magpie.
Grasshopper Warbler.[1]	Turtle Dove.
Great Titmouse.	Wood Pigeon.
Red-throated Pipit.	

He confirms, moreover, Mr. Rowley's remark that the Cuckoo's egg is occasionally found in the nest of the Brambling (*Fringilla montifringilla*).

I have still to name four species which are not included in any of the above lists, *viz.*, the Spotted Flycatcher, Yellow Wagtail, Grey Wagtail, and Wheatear. They were noticed by me some years ago in the first work I ever published.[2] In the case of the Wheatear, a nest of that bird containing three eggs of the Wheatear and one of the Cuckoo was placed under a

---

[1] This species, however, is included in Dr. Thienemann's list above given.

[2] "The Birds of Middlesex," 1866, p. 120.

clod, and in such a position as strongly to favour the opinion of some naturalists that the Cuckoo first lays her eggs and then deposits them with her bill in the nest.

Considering the amount of attention which has been bestowed upon the Cuckoo by naturalists in every age down to the present, one would suppose that every fact in connection with its life-history was now pretty generally known. Such, however, is not the case. There are still certain points which require investigation, and which, owing chiefly to the vagrant habits of the bird, are not easily determined.

How can it be ascertained with certainty, for example, whether the same hen Cuckoo always lays eggs of the same colour, or whether (admitting this to be the case) she invariably lays in the nest of the same species—that is, in the nest of that species whose eggs most nearly approximate in colour to her own?

And yet we must be satisfied on these points if we are to accept the ingenious theory of Dr. Baldamus. If we understand the learned

German rightly, he states that, with a view to insure the preservation of species which would otherwise be exposed to danger, Nature has endowed every hen Cuckoo with the faculty of laying eggs similar in colour to those of the species in whose nest she lays, in order that they may be less easily detected by the foster parents, and that she only makes use of the nest of some other species (*i.e.* of one whose eggs do *not* resemble her own) when, at the time she is ready to lay, a nest of the former description is not at hand. This statement, which concludes a long and interesting article on the subject in the German ornithological journal "Naumannia," for 1853, has deservedly attracted much attention. English readers were presented with an epitome of this article by Mr. Dawson Rowley in the "Ibis" for 1865, and the Rev. A. C. Smith, after bringing it to the notice of the Wiltshire Archæological Society in the same year, published a literal translation of it in the "Zoologist" for 1868. More recently, an article on the subject, by Professor Newton, appeared in "Nature" and elicited various critical remarks from Mr. H. E. Dresser, Mr. Layard, and other ornithologists which deserve perusal.[1]

To enter fully upon the details of this interesting subject would require more space than can here be accorded; one can only glance therefore at the general opinions which have been expressed in connection with it.

If the theory of Dr. Baldamus be correct, is it possible to give a reasonable and satisfactory explanation of it? This question has been answered by Professor Newton in the article to which we have just referred. He says:—
"Without attributing any wonderful sagacity to the Cuckoo, it does seem likely that the bird which once successfully deposited her eggs in a Reed Wren's or a Titlark's nest, should again seek for another Reed Wren's, or a Titlark's nest (as the case may be) when she had an egg to dispose of, and that she should continue her practice from one season to another. We know that year after year the same migratory bird will

---

[1] *See* "Nature," 18th Nov. and 23rd Dec., 1869, 6th Jan., 7th July, and 18th Aug. 1870.

return to the same locality, and build its nest in almost the same spot. Though the Cuckoo be somewhat of a vagrant, there is no improbability of her being subject to thus much regularity of habit, and indeed such has been asserted as an observed fact. If, then, this be so, there is every probability of her offspring inheriting the same habit, and the daughter of a Cuckoo which always placed her egg in a Reed Wren's or a Titlark's nest doing the like." In other words, the habit of depositing an egg in the nest of a particular species of bird is likely to become hereditary.

This would be an excellent argument in support of the theory, were it not for one expression, upon which the whole value of the argument seems to me to depend. What is meant by the expression "once successfully deposited"? Does the Cuckoo ever revisit a nest in which she has placed an egg, and satisfy herself that her offspring is hatched and cared for? If not (and I believe such an event is not usual, if indeed it has ever been known to occur), then nothing has been gained by the selection of a Reed Wren's or Titlark's nest (as the case may be), and the Cuckoo can have no reason for continuing the practice of using the same kind of nest from one season to another.

While admitting therefore the tendency which certain habits have to become hereditary in certain animals, I feel compelled to reject the application of this principle in the case of the Cuckoo, on the ground that it can only hold good where the habit results in an advantage to the species, and in the present instance we have no proof either that there is an advantage, or, if there is, that the Cuckoo is sensible of it.

Touching the question of similarity between eggs laid by the same bird, Professor Newton says:—"I am in a position to maintain positively that there is a family likeness between the eggs laid by the same bird" (not a Cuckoo) "even at an interval of many years," and he instances cases of certain Golden Eagles which came under his own observation. But do we not as frequently meet with instances in which eggs laid by the same bird are totally different in appearance? Take the case of a bird which lays four or five eggs in its own nest before it commences to sit upon them—for example, the Sparrow-Hawk, Blackbird, Missel-Thrush, Carrion Crow, Stone Curlew, or Black-headed Gull. Who has not found nests of any or all of these in which one egg, and sometimes more, differed entirely from the rest? And yet in each instance these were laid, as we may presume, not only by the same hen, but by the same hen *under the same conditions*, which can be seldom, if ever, the case with a Cuckoo.

Looking to the many instances in which eggs laid by the same bird, in the same nest, and under the same circumstances, vary *inter se*, it is not reasonable to suppose that eggs of the same Cuckoo deposited in different nests, under different circumstances, and, presumably, different conditions of the ovary, would resemble each other. On the contrary, there is reason to expect they would be dissimilar. Further, I can confirm the statement of Mr. Dawson Rowley, who says:[1] "I have found two types of Cuckoo's eggs, laid, as I am nearly sure, by the same bird."

---

[1] "Ibis," 1865, p. 183.

It is undeniable that strong impressions upon the sense of sight, affecting the parent during conception or an early stage of pregnancy, may and do influence the formation of the embryo, and it has consequently been asserted that the sight of the eggs lying in the nest has such an influence on the hen Cuckoo, that her egg, which is ready to be laid, assumes the colour and markings of those before her. This is not, however, supported by facts, for the egg of a Cuckoo is frequently found with eggs which do not in the least resemble it (*e.g.* those of the Hedge-Sparrow); or with eggs which, from the nature of the nest, could not have been seen by the Cuckoo (as in the case of the Redstart, Wren, or Willow Wren); or deposited in a nest before a single egg had been laid therein by the rightful owner. Again, two Cuckoo's eggs of a different colour have been found in the same nest. If both were laid by one bird, we have a proof that the same Cuckoo does not always lay eggs of the same colour; if laid by different birds, then the Cuckoo is not so impressionable as has been supposed.

What really takes place, I believe, is this :—

The Cuckoo lays her egg upon the ground; the colour of the egg is variable according to the condition of the ovary, which depends upon the age of the bird, the nature of its food, and state of health at the time of oviposition. With her egg in her bill, the bird then seeks a nest wherein to place it. I am not unwilling to accept the suggestion that, being cognizant of colour, she prefers a nest which contains eggs similar to her own, in order that the latter may be less easily discovered by the foster parents. At the same time the egg in question is so frequently found amongst others which differ totally from it in colour, that I cannot think the Cuckoo is so particular in her choice as Dr. Baldamus would have us believe.

The manner in which "the cuckowe's bird useth the sparrow," "oppressing his nest," living upon him, and finally turning him adrift, has furnished a theme for poets and prose writers in all ages, and has awakened in no small degree the speculative powers of naturalists.

The story is as old as the hills, and it would probably be difficult, if not impossible, to trace it to its origin. It was known to the ancients that the Cuckoo leaves its eggs to be hatched by other birds, but they mingled fact with fable, believing, or at all events asserting, that the young Cuckoo devoured not only its foster brothers and sisters, but ultimately its foster parents. Hence the expression which Shakespeare put into the mouth of the Earl of Worcester to the effect that the youngster

> " Grew by our feeding to so great a bulk
> That even our love durst not come near his sight
> For fear of swallowing."—*Henry IV.* act v. sc. 1.

But though so time-worn is the tale as to be very generally believed, it is singular how few writers have attempted to show a foundation for it from their own observations. So scattered, indeed, is the evidence on the subject, that many naturalists of the present day still hesitate to believe the story, pronouncing the alleged feat of strength on the part of the young Cuckoo to be "a physical impossibility."

Although my present purpose is to direct attention to the latest observations upon this vexed question which have come to us with authority, it will not be superfluous to glance very briefly at what had already been advanced in support of the statement referred to.

Dr. Jenner says positively ("Phil. Trans." vol. lxxviii. p. 225):—" I discovered the young Cuckoo, though so newly hatched, *in the act* of turning out the young Hedge-Sparrow. The little animal, with the assistance of its rump and wings, contrived to get the bird upon its back, and making a lodgement for its burden by elevating its elbows, clambered backwards with it up the side of the nest till it reached the top, where, resting for a moment, it threw off its load with a jerk, and quite disengaged it from the nest. It remained in the situation for a short time, feeling about with the extremities of its wings, as if to be convinced whether the business was properly executed, and then dropped into the nest again."

Montagu, in the Introduction to his "Ornithological Dictionary," states that he took home a young Cuckoo five or six days old, when, to use his own words: " I frequently saw it throw out a young Swallow (which was put in for the purpose of experiment) for four or five days after. This singular action was performed by insinuating itself under the Swallow, and with its rump forcing it out of the nest with a sort of jerk. Sometimes, indeed, it failed after much struggle, by reason of the strength of the Swallow, which was nearly full feathered; but, after a small respite from the seeming fatigue, it renewed its efforts, and seemed continually restless till it succeeded."

Mr. Blackwall, who published some observations on this point in the fourth volume of the " Manchester Memoirs " (second series), says that a nestling Cuckoo, while in his possession, turned both young birds and eggs out of its nest, in which he had placed them for the purpose. He further observed " that this bird, though so young, threw itself backwards with considerable force when anything touched it unexpectedly," an observation subsequently confirmed by Mr. Durham Weir in a letter to Macgillivray.[1]

---

[1] "Hist. Brit. Birds," vol. iii. p. 128.

Mr. Weir says a young Cuckoo was hatched with three young Titlarks on the 6th June. "On the afternoon of the 10th two of the Titlarks were found lying dead at the bottom of the ditch; the other one had disappeared." Subsequently this Cuckoo was removed, and placed in another Titlark's nest, nearer home, for more convenient observation. On the following day Mr. Weir found it covered by the old Titlark "with outstretched wings from a very heavy shower of rain * * * while her own young ones had in the meantime been expelled by the Cuckoo, and were lying lifeless within two inches of her nest." Another instance is given wherein two Cuckoos were hatched in a Titlark's nest. "On the third or fourth day after this the young Titlarks were found lying dead on the ground, and the Cuckoos were in possession of the nest." Ultimately one of the latter, the weaker of the two, disappeared.

A German naturalist, Adolf Müller, of Gladenbach, writing in a German periodical, "Der Zoologische Garten," in October, 1868, has given a curious account of the conduct of two young Cuckoos, which were hatched in the nest of a Robin. A translation of this account was published in "The Field" of Nov. 21, 1868, and it will be unnecessary therefore to give more than the merest outline of the facts detailed in it.

Two young Cuckoos, five or six days old, were found in a Robin's nest, four Robin's eggs lying on the heath before the nest. The two birds were extremely restless, striving to push each other out of the nest, the smaller one always the more active. Herr Müller placed the smaller on the back of the larger one, which immediately began to heave it upwards, and, thrusting its claws into the moss and texture of the nest, actually succeeded in pushing it to the edge of the nest and about four inches further amongst the heath stems. After every contest which was observed both birds contrived to creep back again into the nest. Ultimately the larger one was found lying dead outside the nest, while the Robin was sitting on the smaller bird and the eggs, which had been replaced.

The latest contribution on the subject is that of Mr. Gould, who in his splendid folio work on " The Birds of Great Britain," expressed himself a disbeliever in the popular story. He has since found reason to change his opinion, for in his recently published octavo " Introduction " to that work he says : " I now find that the opinion ventured in my account of this species as to the impossibility of the young Cuckoo ejecting the young of its foster parents at the early age of three or four days is erroneous ; for a lady of undoubted veracity and considerable ability as an observer of nature, and as an artist, has actually seen the act performed [he seems to overlook the circumstance that others had previously seen it], and has illustrated her statement of the fact by a sketch taken at the time, a tracing of which has been kindly sent to me."

This tracing he has reproduced as an engraving in the " Introduction " referred to, and as he has been good enough to allow me the use of the wood block to illustrate the present remarks, the reader may consider himself in possession of a fac-simile sketch from nature.

The following is the account given by Mrs. Blackburn (the lady referred to) of the circumstance as it came under her observation :[1]—

" The nest which we watched last June, after finding the Cuckoo's egg in it, was that of the Common Meadow Pipit (Titlark, or Moss-Cheeper), and had two Pipit's eggs, besides that of the Cuckoo.

" It was below a heather bush, on the declivity of a low abrupt bank, on a Highland hill-side, in Moidart. At one visit the Pipits were found to be hatched, but not the Cuckoo.

" At the next visit, which was after an interval of forty-eight hours, we found the young Cuckoo alone in the nest, and both the young Pipits lying down the bank, about ten inches from the margin of the nest, but quite lively after being warmed in the hand. They were replaced in

---

[1] It would seem that this account was first published by Mrs. Blackburn, in what she terms "a little versified tale of mine," entitled " The Pipits," which appeared in Glasgow in 1872.

the nest beside the Cuckoo, which struggled about till it got its back under one of them, when it climbed backwards directly up the open side of the nest, and hitched the Pipit from its back on to the edge. It then stood quite up-

right on its legs, which were straddled wide apart, with the claws firmly fixed half-way down the inside of the nest among the interlacing fibres of which the nest was woven; and, stretching its wings apart and backwards, it elbowed the Pipit fairly over the margin so far that its struggles took it down the bank instead of back into the nest.

"After this the Cuckoo stood a minute or two, feeling back with its wings, as if to make sure that the Pipit was fairly overboard, and then subsided into the bottom of the nest.

"As it was getting late, and the Cuckoo did not immediately set to work on the other nestling, I replaced the ejected one and went home. On returning next day both nestlings were found dead and cold, out of the nest. I replaced one of them, but the Cuckoo made no effort to get under and eject it, but settled itself contentedly on the top of it. All this I find accords accurately with Jenner's description of what he saw. But what struck me most was this: The Cuckoo was perfectly naked, without a vestige of a feather, or even a hint of future feathers; its eyes were not yet opened, and its neck seemed too weak to support the weight of its head. The Pipits had well-developed quills on the wings and back, and had bright eyes, partially open; yet they seemed quite helpless under the manipulations of the Cuckoo, which looked a much less developed creature. The Cuckoo's legs, however, seemed very muscular; and it appeared to feel about with its wings, which were absolutely featherless, as with hands, the 'spurious wing' (unusually large in proportion), looking like a spread-out thumb. The most singular thing of all was the direct purpose with which the blind little monster made for the open side of the nest, the only part where it could throw its burthen down the bank."

Notwithstanding the objections put forward by sceptics, it is impossible, after reading the evidence of the above-named independent observers, to doubt that the young Cuckoo is capable of doing all that has been attributed to it in the way of ejectment. But it is still very desirable that some competent anatomist should examine and report upon the arrangement and development of the nerves and muscles, which must differ very considerably from those which are to be found at the same age in the young of other insessorial birds.

THE WRYNECK.

(*Jynx torquilla.*)

FOLLOWING closely in the wake of the Cuckoo, if not occasionally preceding it, comes the Wryneck, or Cuckoo's-mate, as it is popularly called from the habit referred to. In some respects it is a very remarkable bird, for not only is its appearance quite unlike that of any other of our summer migrants, but its actions and habits are also totally different. In size no larger than a Skylark, it at once attracts attention by the beauty of its plumage which, although of sombre hue, is prettily variegated with greys and browns of different shades, here and there relieved with black. The under parts, of a soft grey inclining to yellow, are transversely bound with delicate wavy lines. Although for the purpose of comparison, this species may be likened in point of size to the familiar Lark, its structure and habits fit it for a very different mode of life. It is a scansorial or climbing bird, like the Woodpeckers, with toes directed two in front and two behind; hence the term yoke-footed, which has been applied to the particular group of birds in which it is included. The genus to which this bird belongs has generally been associated with the genus *Picus*, to which it undoubtedly bears some affinity. The extensibility of the tongue is the chief character which they have in common, but in the one the extremity is barbed, in the other it is smooth. The fourth toe in the Woodpecker is directed somewhat outwards and backwards, whereas in the Wryneck its natural position is directly backwards, parallel to the first. The bill of the latter more nearly resembles that of *Picus* than that of *Cuculus*, although it is not wedge-shaped at the point. On the other hand the tail has no resemblance to that of the Woodpecker. The genus *Jynx*, therefore, seems to stand between these two genera and to form as it were their connecting link.

The colour of the plumage so closely assimilates to that of the bark and boughs of trees, that it is often difficult to detect the bird when in close proximity to such surroundings. But although the Wryneck may be considered as strictly a woodland bird, adapted by its peculiar structure to climbing the boles of trees and probing the interstices of the bark for lurking insects, it nevertheless finds a considerable portion of its food on the ground, and it especially affects the neighbourhood of ant-hills, where it preys largely on those insects and their larvæ. In this employment its remarkable tongue, like that of the Woodpecker's, is of great service. It is long and slender, with a horny point, and is capable of being protruded for more than twice the length of the head, in consequence of the extreme elongation of the two branches of the flexible or hyoid bone, as it is termed, which supports the tongue, curling round at the back of the head, dividing and passing over each eye, at the forehead, where the branches reunite and extend to the base of the upper mandible. Two long salivary glands, situated beneath the tongue, open into the mouth by two ducts, and secrete a viscid fluid which covers the tongue, and thus causes ants, larvæ, and other small insects forming the food of this species to adhere to it. Where the soil is loose the tongue is thrust into all the crevices to rouse the ants, and for this purpose the horny extremity is very serviceable as a guide to the tongue. The peculiar habit which the bird has of twisting the neck with a slow undulatory movement, like that of a snake, has obtained for it the name of Wryneck, not only in England but throughout the continent, wherever the bird is known.

Although common in the southern and southeastern counties of England, the Wryneck is only partially distributed in the British Islands, and the limit of its geographical area is almost coincident with that of the Nightingale before noticed. In the western and northern counties of England, as well as in Wales, it is comparatively a scarce bird; in Scotland it is very rare, and in Ireland quite unknown. Its arrival in April is speedily announced by its loud and oft-repeated cry, which has been likened to the syllable—" dear, dear, dear, dear, dear," and which resembles, though less harsh, the cry of the Kestrel.

In its mode of nidification, the Wryneck resembles the Woodpeckers, selecting a hole in a tree wherein to deposit its eggs, which are six or seven in number, pure white, and laid with little or no attempt at a nest upon chips of decayed wood at the bottom of the hole. In about three weeks the young are hatched, and both parents take their turn at feeding them, bringing ants and their eggs in mouthfuls, woodlice, small spiders, and other insects. The ants, which are gathered up wholesale by means of the long glutinous tongue, which the bird darts amongst them with great rapidity, are stored up in the mouth until the return to the nest, when they are ejected in ball-like masses into the open gapes of the clamorous young. The latter quickly assume their feathers, and by the month of August are ready to leave the country with their parents *en route* for Africa, Asia Minor, and India, where they pass the cold months of the year. But according to the observations of Lindermayer, Dr. Kruper, and others, many spend the winter in Greece amongst the olive groves, and Lord Lilford has seen it in Epirus in March and December ("Ibis," 1860, p. 235).

In Tangiers, Tripoli, Algeria, Egypt, Nubia, and Abyssinia, it is by no means uncommon. It occurs also in Arabia, and according to Dr. Jerdon ("Birds of India," vol. i. p. 303) is found throughout India, except, perhaps, on the Malabar Coast, where he never saw it, or heard of a specimen being procured. He adds, "It is chiefly, perhaps, a cold-weather visitant in the south of India; but it is found to remain all the year further north."

I have already touched upon the question whether any of our summer migrants breed in their winter quarters, as well as in their summer haunts (see p. 41), and it may be well to note here the above remark of Dr. Jerdon, as well as the observation of Captain Loche, that the Wryneck breeds in the forests of Algeria. It of course remains to be shown whether the individuals which rear their young south of the Mediterranean, ever migrate into Europe; for it is possible that Algeria may be the northernmost limit in summer of those birds which have passed the winter many degrees further south than have the migrants from Europe.

THE HOOPOE.

(Upupa epops.)

AMONGST the large number of migratory birds which resort to the British Islands in spring for the purpose of nidification, are a few which come to us accidentally, as it were, or as stragglers from the main body of immigrants which, crossing the Mediterranean from Africa, becomes dispersed over the greater part of Europe. The Hoopoe is one of these. Not a summer elapses without the appearance, and, I regret to say, the destruction, of several of these beautiful birds being chronicled in some one or other of the many periodicals devoted to Natural History. If the thoughtless persons, whose first impulse on seeing an uncommon bird, is to procure a gun and shoot it, would only take as much pains to afford it protection for a time, observe its habits, describe its mode of nesting and manner of feeding its young, they would do a much greater service to ornithology by recording the result of their observations, than by publishing the details of a wanton destruction.

That the Hoopoe will breed in this country, if unmolested, is evidenced by the recorded instances in which it has done so where sufficient protection has been afforded it during the nesting season. Montagu states, in his " Ornithological Dictionary," that a pair of Hoopoes began a nest in Hampshire, and Dr. Latham has described a young Hoopoe which was brought to him in the month of June. A pair frequented Gilbert White's garden at Selborne;

and another pair nested for several years in the grounds of Pennsylvania Castle, Portland.[1] Mr. Jesse states[2] that some years ago a pair of Hoopoes built their nest and hatched their young in a tree close to the house at Park End, near Chichester; and according to the observations of Mr. Turner, of Sherborne, Dorsetshire, the nest has been taken, on three or four occasions, by the schoolboys from pollard willows on the banks of the river Lenthay. The birds were known to the boys as "Hoops."

In the same county, on the authority of the Rev. O. P. Cambridge, a pair of Hoopoes are reported to have bred at Warmwell. The Rev. A. C. Smith, of Calne, Wilts, says that a nest, containing young birds, was taken many years ago in his neighbourhood; and another nest, according to Mr. A. E. Knox, was found at Southwick, near Shoreham. Canon Tristram states that the Hoopoe has bred at least on one occasion, in Northamptonshire.

---

[1] *Cf.* Garland, "Naturalist," 1852, p. 82.
[2] "Gleanings in Natural History."

Mr. Howard Saunders informs me that many years ago a pair of Hoopoes took possession of a hole in a yew tree in the shrubbery of a garden at Leatherhead, and reared their young in safety. He afterwards saw both old and young birds strutting about on the lawn. I have seldom met with this bird in England, and then only on the coast in September, when the beauty of its plumage had become faded, and the feathers ragged, and it was about to emigrate southwards for the winter. But on the continent, and more particularly in France, I have had many opportunities of observing it, and noting its actions and habits. In its movements on the ground it struck me as resembling the Rook more than any other bird I could think of at the time; the same stately tread and gentle nodding of the head, every now and then stopping to pick up something. It does not carry the crest erect, but inclining backwards, and is less sprightly in its movements generally than I had previously supposed. On the wing it at first sight reminds one of a Jay, the principal colours being the same, viz., black, white, and pale cinnamon brown ; but the distribution of colour is different, and the flight is not so rapid, and more undulating. The wings are large for the size of the bird, and the first-quill feather being much shorter than the second, the wing has a rounded appearance which makes the flight seem heavier.

It is a shy bird, taking wing on the least alarm, except when surprised by a hawk or other large bird, when, according to the observations of the German naturalists, Naumann and Bechstein, it resorts to a very singular expedient to protect itself. It squats upon the ground, spreads out its tail and wings to their fullest extent, bringing the primaries round so as almost to meet in front, and throws back its head and bill, which it holds up perpendicularly.[1] So long as danger threatens, it remains in this

---

[1] For a notice of this singular habit I am indebted to my friend Mr. H. E. Dresser, who has translated Naumann's observations on the subject in his beautiful work on the "Birds of Europe."

odd position, probably to deceive the enemy; for when thus spread out, at a little distance it looks more like an old parti-coloured rag than a living bird.

The Hoopoe lives a good deal on the ground where it finds its chief food, which consists of beetles of various kinds, and their larvæ, caterpillars, and ants. It is especially partial to dung-beetles, and may often be seen in search of them upon the roads, where it is also fond of dusting after the manner of a Skylark. But besides picking up a great deal of food from the surface, it also probes beneath the soil where the nature of the ground admits of this, and secures many a worm and lurking grub by means of its long and slender pointed bill. It swallows a beetle or other small morsel just as the Hornbills in the Zoological Society's Gardens swallow the grapes which are thrown to them, that is to say, it seizes it first between the tips of the mandibles, then throwing the head back suddenly, and opening the bill at the same instant, the food is jerked into the gullet with great precision, and disappears. When it seizes a worm, however, the process is somewhat different. It bruises it by beating it against the ground, pinches it all over between the mandibles, and finally swallows it lengthwise with sundry jerks of the head.

In other respects, as well as in the mode of taking their food, the Hoopoes resemble the Hornbills. They build in holes of trees as the latter are known to do, and the hens sit upon the eggs without interruption until they are hatched, the males, as in the case of the Hornbills, bringing food and feeding them from the outside of the hole. The eggs, which are generally five or six in number, are elongated, nearly oval, and of a greenish grey colour. The young when first hatched are naked, but soon get covered with small blue quills from which the feathers sprout. They are unable to stand upright until nearly fledged, but crouch forward and utter a hissing noise. Their crests are soon developed, but their bills do not acquire their full length until the following year.

Lord Lilford states that although the Hoopoe as a rule prefers a hole in an old ash or willow tree for nesting in, he has seen a nest on the ground under a large stone, others in holes on the sunny side of mud or brick walls, one in a fissure of limestone rock, and another in a small cavern.

Dr. Carl Bolle has observed that in the Canaries, where trees are scarce, the Hoopoe breeds in holes of the stone walls and clefts of the rocks.

During his residence in China, where this bird is common, Mr. Swinhoe was surprised to find that it often breeds in the holes of exposed Chinese coffins, whence the natives have a great aversion to them, branding them as "Coffin-birds;" and the Russian naturalist Pallas once found a nest of the Hoopoe, containing seven young ones nearly ready to fly, in the decomposed abdominal cavity of a dead body!

The note of the Hoopoe is very remarkable, and not to be mistaken for that of any other bird with which I am acquainted. It sounds like the syllables "hoop-hoop," "hoop-hoop," frequently repeated, and in the quality of its tone approximates to the call of the Cuckoo, but the second note is a repetition of the first instead of being, as in the case of the Cuckoo, a third below it. Old authors affirmed that this peculiar sound was produced by the bird distending its cheeks with air, and tapping its bill upon the ground, thereby causing the notes to escape as it were spasmodically. This curious statement has received some confirmation from the observations of Mr. Swinhoe.[1] He says: "To produce these notes, the bird draws the air into its trachea, which puffs out on either side of the neck, and the end of the bill is tapped perpendicularly against a stone or the trunk of a tree, when the breath being forced down the tubular bill produces the correct sound." He adds, however, that he has observed a Hoopoe perched upon a hanging rope, and uttering its well-known cry without any tapping of the bill.

---

[1] *Cf.* "Zoologist," 1858, and "Proc. Zool. Soc.," 1863, p. 264.

I cannot help thinking that a bird observed in the act of calling *whilst picking up food*, as many species do, has given rise to the notion that the sound is *produced* by tapping, whereas in truth it precedes and follows the movement. The only motion that I could ever detect in a Hoopoe whilst calling was a nodding of the head, and a depression of the crest-feathers.

From the accounts which have been handed down to us by old authors, and the numerous specimens which may be seen preserved in old collections, it would appear that the Hoopoe was formerly much more plentiful in England than it is at the present day. The decrease in its numbers probably arises from two causes, viz., the clearance of forest land, entailing the destruction of many old trees which were once attractive as nesting places,[1] and the increased use of fire-arms which unfortunately results in

---

[1] Mr. Benzon of Copenhagen informed my friend Mr. Dresser that a short time ago the Hoopoe was by no means rare in Norway, but now that the forests have been cleared of all the old and hollow trees it has entirely vanished from the fauna of his district.

the destruction of many of these beautiful birds, at a time when they are just about to pair and commence nidification.

The period of its migration into Europe in the spring sets in early in April. The late Commander Sperling, when stationed with his vessel in the Mediterranean, frequently met with Hoopoes at sea during their passage. In the English Channel on the 15th April, 1854, a Hoopoe after flying two or three times round a steamer entered one of the windows of the saloon and was taken, apparently exhausted with fatigue. Another, on the 21st April, alighted on a mackerel-boat between the Eddystone Lighthouse and Plymouth Breakwater, in an exhausted state, and allowed itself to be taken.

The average date of arrival in England may be said to be the third week in April, when the species is more frequently met with in the eastern and south-eastern counties, although it wanders inland to a considerable distance. It is regarded by Mr. R. Gray[1] as a straggler to

---

[1] "The Birds of the West of Scotland," p. 198.

Scotland; and Mr. Thompson remarks[1] that in Ireland it has appeared occasionally in all quarters of the island.

As autumn approaches, these birds, or such of them as have contrived to escape destruction, begin to move southwards for the winter, and passing gradually down to the Mediterranean, are observed for some days about the groves and olive gardens near the sea before they finally cross over. In this way they return to their winter haunts about the end of August or beginning of September. Throughout Southern and South-eastern Europe, as well as in Siberia and North-eastern Africa, the Hoopoe breeds commonly; but in the northern and western parts of the last-named continent it is chiefly a winter visitant. The Siberian birds, probably, and not the European ones, migrate to India and China for the cold season, and some remain to breed in both these countries. Those which have passed the summer in Europe, as already shown, spend their winter in Africa.

---

[1] "Nat. Hist. Ireland" (Birds), vol. i. p. 353.

Occasionally a Hoopoe has been observed in winter in the British Islands, but so rarely as to make the occurrence a matter of note. An instance or two of this kind in Norfolk has been noticed by Hunt in his "British Birds" (vol. ii. p. 147); and Mr. R. Gray, in his "Birds of the West of Scotland," p. 198, refers to two which were killed near Glasgow, in different years, so late as the month of October.

The late Sir William Jardine informed me that two were shot in Dumfriesshire in the winter of 1870-71.

The most perfect specimen of the Hoopoe I have ever seen is one in my collection, which was shot at the Dell, a piece of water near Whetstone, Middlesex, on the 25th April, 1852. It has no less than twenty-two crest feathers the longest two inches in length, arranged in two parallel rows, with the upper surfaces outwards, and of a pale cinnamon colour broadly tipped with black. The other portions of the plumage are equally perfect and bright in colour.

THE GOLDEN ORIOLE.

(Oriolus galbula.)

LIKE the Hoopoe, the Golden Oriole makes its annual visit to the European continent from the countries south of the Mediterranean, in the month of April, and returns in September. In the interval it may be found not uncommonly in the wooded parts of Central and Southern Europe; but it is rare in the north, being seldom seen in Sweden, and unknown in Norway.

In England, where it may be regarded as an

irregular summer migrant, it unfortunately meets with little or no protection, for its bright colours at once attract attention, and many get shot before they have been a week on our shores. The male bird is bright yellow, with black wings and a black and yellow tail. The female is dull green, with pitch-brown wings, the upper tail coverts greenish yellow, and the under parts greyish white, longitudinally streaked with brown on the shafts of the feathers; the flanks yellow, and streaked in the same way.

My impressions on meeting with Golden Orioles for the first time in France, now many years ago, will not be easily forgotten. I wanted to see them alive, hear their notes, shoot two or three to examine them closely, and ascertain the nature of their food; and accordingly I accepted the invitation of a friend and took up my quarters at an old country house, about halfway between Paris and Orleans. On looking over my note-book for that particular year, I find the following entry, relating to the Golden Oriole :—

"Long before six in the morning I was

awakened by a perfect chorus of birds—Blackcap, Nightingale, Thrush, Wood Pigeon, Chaffinch, Starling, and Magpie were all recognized; but what pleased me above all, was a beautiful mellow whistle, which I took to be that of the Golden Oriole, and in less than an hour afterwards I found that I was right in my surmise, for on walking through the woods which flank one side of the house, I had the pleasure of seeing for the first time alive several of these beautiful birds. They were very shy, and kept to the tops of the oak trees; but by proceeding cautiously I managed to get near enough to see and hear them well. Their note is really splendid, so mellow, loud, and clear—something of the Blackbird's tone about it, but yet very different; while in their mode of flight and perching they remind one of a Thrush. After a long search, I at length found a nest, placed at the extremity of a thin bough, and at the top of an oak tree, about sixty feet up. There were no branches for more than thirty feet, and it would have been almost impossible to reach it without assistance. I therefore marked the spot, and determined to get a long ladder a little later and try and take it. The keeper informed me that it was early yet for Orioles' eggs, and so I left the nest for the last day of my stay here. In the afternoon I went with the keeper to the Parc de Marolles. We could hear the Orioles, or *Loriots*, as the French call them from their notes, singing loudly in the recesses of the woods; but the foliage was so thick, and they kept so much to the tops of the trees, that it was almost impossible to catch sight of them. Their greenish-yellow feathers, too, harmonized so well with the leaves, that it rendered them still more difficult to see.

"Following the direction of the notes, I continued to make my way through the underwood as noiselessly as possible, peering through the branches, and striving in vain to catch sight of a bird. For a long time the sound seemed to be as far away as ever, or, as I advanced it receded. The sun was broiling hot, and the exertion of forcing my way through the underwood, and straining my neck forward in my endeavours to get a sight of the bird, put me in a profuse perspiration. The result of about three hours' work was, that I finally succeeded in getting three shots at long intervals, and secured a pair of Orioles, a young male and an old female. Subsequently, however, I got others. I found the stomachs of these birds crammed with caterpillars of various species, and can well understand the good they do in young plantations, by ridding the trees of these pests.

"The colours of the soft parts in these birds, as noted by me at the time, were as follows:— Iris, reddish hazel; bill, brownish flesh colour; legs and toes, pale lead colour.

"On June 3rd, after breakfast, I went to the wood near the house to take a Golden Oriole's nest, and a difficult matter it was. The nest was placed in a slender fork at the extremity of a thin bough of an oak tree, and almost at the top.

"The oaks here are not, as in England, sturdy

and short, with wide-spreading heads, but tall and slender, running up for a great height without any branches, and very tiring to climb. I was obliged to saw off the branch before I could look into the nest, and after a great deal of trouble, when I at length got it down safely, I found, to my disappointment, that it contained three young birds instead of eggs. Could I have ascertained this without cutting off the branch, I should certainly have left them where they were; as it was, there was no help for it but to take them. They were apparently about three days old, and almost naked, the skin of an orange or yellowish flesh-colour very sparsely flecked with yellow down. I fed them on maggots, and covered them with cotton wool to keep them warm, and in this way I kept them alive until I reached Paris, where they died, and were entrusted to a skilful taxidermist for preservation."

Although the discovery of a Golden Oriole's nest in England is not unprecedented, it is of sufficiently rare occurrence to attract the attention of naturalists, more especially when the finder (as in the case to which I am about to allude) has the humanity and good sense to permit the young to be reared, instead of shooting the parent birds the moment they are discovered, and thus effectually putting a stop to all attempts at nidification.

It is a pleasure to be able to record the fact, that during the summer of 1874, a pair of Golden Orioles took up their quarters in Dumpton Park, Isle of Thanet, where—the proprietor, Mr. Bankes Tomlin, having given strict injunctions that they should not be disturbed—they built a nest, and successfully reared their young, ultimately leading them away in safety.

They must have commenced building somewhat later than usual, for it was not until the 6th of July that I first heard of the nest, and the young were then just hatched. Mr. Bankes Tomlin having kindly invited me to come and see it, I lost no time in availing myself of his invitation, and a few days later, namely, on

July 12th, I found myself at Dumpton Park, standing under the very tree in which the nest was placed. The reader may smile at the idea of journeying from London to Ramsgate merely to look at a nest; but if he be an ornithologist, he will know that Golden Orioles' nests are not to be seen in this country every day, and that when found they are worth " making a note of." Often as I had seen the bird and its nest on the Continent, it had never been my good fortune until then to meet with it in England. Indeed, the instances in which nests of the Oriole have been found here and recorded are so few that they may be easily enumerated. According to the concise account given by Professor Newton in his new edition of "Yarrell's British Birds," one was discovered in June, 1836, in an ash plantation near Ord, from which the young were taken; but, though every care was shown them, they did not long survive their captivity. " Mr. J. B. Ellman says ('Zoologist,' p. 2496) that at the end of May, 1849, a nest was, with the owners, obtained near Elmstone. It was suspended from the extremity of the top branch of an oak, was composed entirely of wool bound together with dried grass, and contained three eggs. Mr. Hulke, in 1851, also recorded ('Zoologist,' p. 3034) a third, of which he was told that it was found about ten years previously in Word Wood, near Sandwich, by a countryman, who took the young, and gave them to his ferrets; and Mr. More, on the authority of Mr. Charles Gordon, mentions one at Elmstead, adding that the bird appeared again in the same locality in 1861. Mr. Howard Saunders and Lord Lilford informed the editor that in the summer of 1871 they each observed, in Surrey and Northamptonshire respectively, a bird of this species, which probably had a nest. Messrs. Sheppard and Whitear speak of a nest said to have been found in a garden near Ormsby, in Norfolk; but the eggs formerly in Mr. Scales's collection, which it has been thought were taken in that county, were really brought from Holland, and the editor is not aware of any collector who can boast the possession of eggs of this species laid in Britain."

The nest which I am now enabled to record was placed in a fork of a very thin bough of an elm tree, at a considerable height from the ground, and almost at the extremity of the branch, so that it was impossible to reach it except by cutting off the branch near the trunk. Happily, in this case there was no need to reach it, and the finder was enabled to ascertain when the young were hatched by sending a man up the tree high enough to look into the nest without disturbing it. A few days before his first ascent there had been a strong wind blowing for some time, and the slender branch was swayed to and fro to such an extent, that, notwithstanding the depth of the saucer-like nest, one of the eggs was jerked out upon the grass below and broken, though not irreparably so. When I saw it, it was in two pieces, but unmistakably the egg of an Oriole—in size equal to that of a Blackbird, but shining white, with black or rather dark claret-coloured spots at the larger end. It has been carefully preserved by Mr. Tomlin.

As long as his man remained in the tree the hen bird continued to fly round, uttering at intervals a loud flute-like note, and occasionally making a curious noise, such as a cat makes when angry.

It is perhaps scarcely necessary to remark that, as regards situation, form, and the materials of which it was composed, the nest did not differ from those which one is accustomed to see on the Continent. Invariably placed in, and suspended under, the fork of a horizontal bough, the sides of the nest are firmly bound to each branch of the fork with blades of dry grasses and fibrous roots. There is generally a good deal of sheep's wool in the nest itself, which, taken in connection with its peculiar shape, gives it a very singular and unique appearance.

On the 12th of July, as we approached the nest in question, the hen bird was sitting, but left as we advanced, and perched in a neighbouring elm, whence at intervals she uttered the peculiar noise to which I have referred. Not wishing to keep her too long from her young, we left the spot in about ten minutes, after carefully inspecting the nest with a binocular. Returning again in half-an-hour, and a third time two or three hours later, we saw the hen on each occasion quit the nest and take up her position, as before, at a little distance. Once only did I catch a glimpse of her more brightly-coloured mate as he darted between two trees. He was very shy, and silent too, being seldom heard, except very early in the morning, or at twilight. This, however, is the case with most song-birds after the young are hatched, for they are then so busy providing food for the little mouths that they have scarcely time to sit and sing. Mr. Tomlin, who had other and better opportunities for observing him, gave me to understand that he was not in the fully adult plumage,[1] so that it seems the males of this

---

[1] On this point the late Mr. Blyth, writing in the Natural History columns of "The Field," 17th August, 1872, under the signature "Z.," remarked that Orioles are amongst the

species breed before they have assumed their beautiful black and yellow colours.

On the 22nd of July the man again ascended the tree and peeped into the nest. The young had flown, but were subsequently discovered sitting about in the park with the old birds. As soon as the nest was no longer wanted, Mr. Tomlin had the branch which supported it cut off, and, writing to me on the subject the following day, he observed, that " upon examining the nest we found the corners tightly bound with long pieces of matting. One would almost imagine that a basketmaker had been at work."

Both the old and young birds continued to haunt the park until the 1st of August, after which date they were no longer seen. The young were, however, well feathered by that

---

few birds which breed before attaining the mature plumage, and the females acquire this later than the males, being always, however, of a greener shade. He had observed this in *O. melanocephalus*, *O. chinensis*, *O. tenuirostris*, and *O. acrorhynchus*, but thought that "the old females of *O. galbula*, and *O. kundoo*, less frequently attain the male colouring than those of the other species mentioned."

time, and able to take care of themselves. Let us hope that they contrived to escape the eyes of prowling gunners beyond the park, and that they will return again in succeeding years to gladden the eyes and ears of their kind protector.

It is much to be wished that other proprietors would follow the good example thus set by Mr. Bankes Tomlin. Could they be induced to do so, they would become acquainted with many beautiful birds which visit us from the Continent every spring, and which would in most cases rear their young here if allowed to remain unmolested. Apart from the gratification to be derived from seeing these brightly-coloured birds within view of the windows, and hearing their mellow flute-like notes, they would be found to be most useful allies to the gardener in ridding the trees of caterpillars, which they devour greedily, and keeping many other noxious insects in check.

THE RED-BACKED SHRIKE.

(*Lanius collurio.*)

QUITE unlike any other of our summer migrants in appearance, the Red-backed Shrike, or Butcher-bird, as it is more frequently called, differs from them all in habits, and from the majority in having no song to recommend it to notice. It is a curious bird in its way, shy and retired in its disposition, and prefers tall tangled hedgerows or the thick foliage of the lower branches of the oak, where it can sit unobservedly and dart out upon its unsuspecting prey. It is a very Hawk by nature, capturing and killing mice, small birds, moths and beetles of every size and description. These when caught are firmly impaled upon the long and strong points of the whitethorn for future consumption, and the odds and ends which may be found thus hung up, as it were, in the Butcherbird's larder are worth notice. On one thorn, perchance, a Blue Titmouse with its head off, on another a small meadow mouse (*Arvicola agrestis*), or perhaps a harvest mouse (*Mus messorius*), on a third a great dor-beetle or a cockchafer, not yet dead, but buzzing round and round upon the sharp thorn, and trying in vain to effect its escape, while above, below, and on all sides may be seen the wingless bodies of large moths, the fluttering forms of dragon-flies, or the remains of beetles.

From this singular habit the bird has earned the name of Butcher-bird, not only in England but in other countries. In France it is termed *l'écorcheur*, the flayer; in Germany it is known as *der Würger* (the strangler, or garotter), and *der Fleischer*, or butcher, whence no doubt is derived " Flusher," the provincial name by which it is known in some parts of England. The Linnæan name for the genus, *Lanius*, has the same signification.

The Red-backed Shrike arrives here somewhat later than most of the summer migrants, and is seldom observed before the first week in May. It is generally found in pairs until after the young are hatched and ready to fly, when the families keep together in little parties until the end of August or beginning of September, when they leave the country.

The note of the Red-backed Shrike resembles the syllables " tst—tst," or " tsook—tsook," loudly uttered, and reminds one a little of the notes of the Whinchat and Stonechat. It has besides a harsh " kurr-r," which it utters when any one approaches the nest, and as it flits from branch to branch, lowering the head, and slowly moving the tail up and down.

The male is decidedly a handsome bird. It has the head and neck grey, with a broad black streak passing from the bill through the eye and ear coverts, the back reddish chestnut, chin white, under parts pale salmon colour, and the wings and tail black, the latter broadly marked with white at the base.

The hen bird is much plainer in appearance, being of a dull and somewhat mottled brown above, and buffy white beneath, with crescentic brown markings on the breast and flanks.

The bill in both is short and thick, the upper mandible hooked at the point and prominently notched or toothed, as in a hawk. The feet are strong, with sharp and curved claws, and well adapted for seizing and holding a struggling prey.

Both birds assist in the construction of the nest, which is a substantial well-built structure of twigs, dry grass, and moss, lined with fibrous roots and horsehair, and is usually placed at some height from the ground in the middle of a whitethorn bush, or thick hedgerow. The eggs, five and sometimes six in number, vary a good deal in colour, being yellowish or greyish white with lilac or pale brown markings disposed in a zone at the larger end, or pale salmon colour, with dull red markings distributed in the same way.

The distribution of this bird in the British Islands is very partial, for it is unknown in Ireland, of rare occurrence in Scotland, and in England is found chiefly in the midland and southern counties. During the summer months it is generally dispersed throughout Europe and the temperate parts of Siberia, and as autumn approaches, it crosses the Mediterranean into Africa, where it travels down the east coast through Egypt, Nubia, and Abyssinia, to Natal, and on the west coast has been met with in Great Namaqua Land, Damara Land, and the Okavango region, where, according to Andersson, it breeds.[1]

Breeding in its winter quarters? Well, that is the question. Can the birds which Andersson found nesting in South-west Africa in our

---

[1] " The Birds of Damara Land," p. 136.

winter, have been the same birds which reared a brood in Europe the previous summer? He says it is migratory in Damara Land. Is the same species, then, found on both sides of the equator, migrating north and south on both sides of it, but never crossing it?

The late Mr. Blyth thought that, with one exception, our summer migratory birds do not breed in their winter quarters, but from what has been recorded of the Swallow, the Sand-martin, the Wryneck, the Turtle-Dove and the present species, there seems room to doubt the correctness of this view.

Another species of Shrike, the Woodchat (*Lanius rutilus*), has been met with in this country during the summer months, and has been reported even to have nested here. It is of extremely rare occurrence, however, and cannot with propriety be included, at least for the present, amongst our annual summer migrants.

THE TURTLE-DOVE.

(Turtur auritus.)

AMIDST the general harmony of the grove in spring, there are few prettier sounds than the gentle cooing of the Turtle-Dove. Perched upon a bough at no great height from the ground, it pours forth its soft murmurings with a delightful *crescendo* and *diminuendo*, while close at hand, upon a mere frame-work of a nest, the mate sits brooding upon her two milk-white eggs.

Arriving in this country towards the end of April or beginning of May, the Turtle-Dove is seen only in pairs until the young are able to fly. Young and old then unite in flocks, and ten or a dozen may often be found together in the pea-fields and on the stubble, where they pick up the fallen grain. They are very partial also to vetches, rape, and wild mustard, and do some service to farmers by ridding the cultivated lands of the seeds of numerous weeds, such as the Corn Spurrey (*Spergula arvensis*), which is common in corn-fields, and the Silver-weed (*Potentilla anserina*), which they find upon the fallows.

When Partridge shooting in September I have frequently found Turtle-Doves feeding amongst the root crops as well as on the bare stubble, but notwithstanding the cover afforded by the turnip-leaves I have generally found them so exceedingly wary, that it required a good deal of manœuvring before I could get a sufficient number to make a pie. In point of flavour, and of course in size, they are not to be compared with the Wood Pigeon, being rather dry and somewhat insipid. Their flight is rapid, and when suddenly flushed they go off at such a pace, that it requires a quick shot to bring one down.

When taken young they are readily tamed, and will even breed in confinement, a thing that rarely happens in the case of the Wood Pigeon. Mr. Stevenson has known two or three instances in which this species when caged has crossed with the Collared Turtle-Dove (*Turtur risoria*) and reared a brood, and others have been recorded. The young "presented many characteristics of both parents."[1]

Although commoner in the eastern and south-eastern counties of England, the Turtle-Dove is generally dispersed in summer throughout the British Islands. In Ireland it is regarded as an annual visitant to the cultivated districts, and it has been found in most of the counties of Scotland, where Mr. Robert Gray, however,

---

[1] "The Birds of Norfolk," vol. i. p. 360.

considers that it can only be ranked as a straggler.[1] All the specimens which have come under his own observation were obtained in spring or autumn. In the Hebrides specimens have been shot in Islay and Skye, but in the outer islands none have been seen. Dr. Saxby says that the Turtle-Dove, " although formerly very scarce in Shetland, may now be seen every year in certain of the gardens—that at Halligarth especially—between spring and autumn. It has always occurred singly. With nearly all the habit was to wander away during the day-time, returning at night to roost in one particular tree.[2]

It was first known to occur in Shetland in the autumn of 1856, when Mr. Edmondston of Buness shot one at Balta Sound. "It was but little seen from that time until about six years ago (1868), by which time the trees had grown above the walls, offering a more suitable refuge for stragglers of this description." On two

---

[1] "The Birds of the West of Scotland," p. 223.
[2] "The Birds of Shetland," p. 152.

occasions the Turtle-Dove has been found in Orkney.[1]

On the continent of Europe this bird seems to be confined chiefly to the central and southern parts, and does not reach Scandinavia or Russia. But in France, Spain, and the countries bordering the Mediterranean, it is very common in summer. Its winter haunts are in North Africa; and in Egypt and Nubia it is especially abundant. Capt. Shelley says that it frequently breeds there.[2] In the neighbourhood of Tangier vast flocks arrive from the interior in April and May to cross the Straits of Gibraltar,[3] and on reaching Andalusia afford considerable diversion to the Spanish sportsmen, who kill large numbers by lying in wait for them.

Mr. Thompson, when proceeding in H. M. S. "Beacon" from Malta to the Morea in the month of April, saw this species on its spring

---

[1] Baikie and Heddle, "Fauna Orcadensis," p. 55, and Gray, *op. cit.* p. 223.

[2] "Birds of Egypt," p. 214. See *ante*, p. 281.

[3] Irby, "Ornithology of the Straits of Gibraltar," p. 134.

migration. One flew on board on the 24th of April, and another on the 25th; they each rested for a short time on the rigging, and then pursued their flight northwards. On the 26th four came from the south, two of them singly, the others in company; one only alighted on the ship; it was caught in the evening when asleep. Throughout the 27th many were observed coming from the south, and generally singly, never more than two together; very few alighted. On the 24th the vessel was at sunset ninety miles east of Sicily, Syracuse being the nearest land; on the 27th, forty-five miles from Zante, and sixty west of the Morea. On the 29th of April one was seen near Navarino; and another on the 6th of May in the island of Syra. At the end of the month numbers were observed amongst the light foliage in the gardens of the old seraglio at Constantinople.

Colonel Irby informs me that when on his way from Southampton to Gibraltar on the 9th October, he saw a Turtle-Dove on its southward migration in the middle of the Bay of Biscay.

THE LANDRAIL OR CORNCRAKE.

(Crex pratensis.)

SPORTSMEN who during the early part of September follow their "birds" into seed—clover, rape, or mustard—seldom fail in such places to pick up a Landrail or two, and add in this way a pleasing variety to their bag. The appearance of a Rail usually gives rise to some comment, and not unfrequently to an expression of surprise that a bird of such skulking habits and apparently of such weak flight should be able to leave the country periodically, and return to it. That it does so, however, is certain.

It arrives here towards the end of April, and remains with us till the end of September. During May and part of June its incessant craking note is constantly reminding us of its presence; and if in July and August its silence has caused us for a time to forget it, we renew acquaintance once more in September, when in quest of nobler and more important game. After that month we may look for it almost in vain, for, although a Landrail now and then turns up during the winter months, its appearance at such times is exceptional, the great majority of these birds having left our shores before the first day of pheasant shooting has come round. We can only account for the appearance of Landrails in winter by supposing them to be individuals of a late brood, unprepared to leave at the proper time, or wounded birds unable to take part in the autumnal migration. In Ireland, however, their occurrence in winter and early spring has been noticed so much more frequently than in England, that a good naturalist there, Mr. Blake Knox, has suggested that the bird hybernates. Writing in the "Zoologist" for 1867, at page 679 he says: "I cannot divest myself of the belief that the Corncrake hybernates, notwithstanding my having found it repeatedly dead in the sea, both during autumn and spring, which many would say should prove migration to the most sceptical. I do not for one moment doubt that it leaves Ireland in numbers in the autumn, but where does it go? Does it hybernate where it goes to? Is it to be met with anywhere in numbers, flying or running, during our winter? Does it only crake in its spring or summer haunts? In support of hybernation, we have the great amount of fat, coming on in winter (Corncrakes often burst from fat when they fall on being shot), which all hybernating animals attain; the number of uninjured and healthy birds found in Ireland during winter, their peculiar skulking habits at this season, the old hollow ditches they frequent, their peculiar apathy and disinclination to fly, and their early appearance without 'craking' (I have seen them in the middle of March) along the sedges of rivers, which would be the first places they would make for after their winter rest. I do not see why hybernation of birds is so much scouted, for scores of animals and millions of insects do so. Many fishes, too, become so torpid that you may fish for weeks and not get one, yet some fine day dozens of the kind you look for will reward your patience; still you have been told or read somewhere that that species migrates from our shores in autumn 'to seek more genial skies,' and that is why they are not caught in winter. The subject is very far from being absurd, though many have considered it equally so with 'Corncrake turning to Water-rail.' I knocked down a ditch bank some years ago in January, and turned out three living Corncrakes, and ate them too. In the year 1861, during November and December, I used frequently to turn out of a particular hole one of these birds; I caught it at last one night in the hole—or nest I might say, for it was thickly bedded with leaves from a neighbouring dunghill, on which beech leaves had been thrown; but I let it go after some time—in honesty, not through kindness, but because I could not help it, for it could pass through any hole, almost, as Paddy used to say, 'as limber as a glove.' I could also state many instances of dogs chasing Corncrakes in winter to holes, and in one case remember how nearly I was summoned for tearing down a man's ditch bank 'in pursuit of rats,' as he said, though he had two eyes and saw the bird run from hole to hole. More learned men than he may have often thought the same thing. Hybernating, in my view, would not mean a dead, torpid state. I should consider it a sleepy, inactive state—a lying-up in cold weather, and a temporary arousing during genial days; and in this state I have met the Corncrake in winter."

Before one can accept the hybernation theory, however, some stronger evidence in its favour would be desirable—the discovery, for instance, of a Landrail in a torpid state from which it might be observed to recover. At present I do not remember to have heard of such an instance, and, so far as one's knowledge of the bird extends, it seems far more probable that it seeks holes in banks and old walls merely for shelter and warmth, in the neighbourhood of which it contrives to find sufficient nourishment to keep it alive, until such time as the increasing warmth of spring brings increase of insects and molluscous food. An examination of the alimentary system of the Water-rail (*Rallus aquaticus*) shows that this bird is no better fitted to withstand an English winter than its cousin the Landrail, and yet it is frequently found by sportsmen upon Snipe-ground at the height of the cold season. Its good condition too at this time testifies to there being a sufficient supply of food, which should be equally obtainable by the Landrail.

The nature of this food is miscellaneous—slugs and snails of several species, small freshwater mollusca, worms, leeches, beetles, the seeds of various weeds, and tips of grass blades; in addition to which the stomach is usually found to contain numerous small particles of gravel or grit as aids to digestion.[1]

In its search for this kind of food, the Landrail must traverse daily an immense tract of ground, for which, however, its strong muscular legs and large feet are well adapted.

For six months at least in the year it appears to be very generally distributed throughout the British Islands; and in Ireland, owing to the more humid climate and the general prevalence of meadow land, it is thought to be even commoner than in England. As regards Scotland, the latest authority on the subject, Mr. Robert Gray, in his " Birds of the West of Scotland," says : " There is, perhaps, no Scottish bird more generally distributed than the familiar Corncrake. It is found in every district, cultivated and uncultivated, on the western mainland, from the Mull of Galloway to Cape Wrath, and also over the whole extent of both groups of islands, and

---

[1] A Landrail caught on Canvey Island, at the mouth of the Thames, lived in confinement on corn and water for a week, when it made its escape.

all the rocky islets on the west coast, extending to Haskeir Rocks, the Monach Islands, and St. Kilda. It will, in fact, take up its abode and rear its young on such places as are almost exclusively frequented by birds dependent on the sea for their daily subsistence, all that can be looked upon as an attraction being but an occasional patch of grass and a moist hollow, to remind it of the distant meadow where, perchance, it had its haunts the previous summer. I have observed it in the uninhabited islands of the Hebridean seas, and have heard it near the summit of Ailsa Craig, rasping its eerie cry after nightfall, as a rude lullaby to the Gulls hatching on the grassy verge of a precipice."

This is by no means the limit of its haunts northward and westward; for besides being found in Denmark, Norway, and Sweden, and the Faroe Isles, it actually visits Greenland, and on several occasions has been met with on the eastern coast of the United States, whither it must have travelled, doubtless, *viâ* Greenland.

A single instance is on record of its having been shot in the Bermudas,[1] although this group of islands is distant from Cape Hatteras—the nearest point of the North American coast—about 600 miles. After this, English sportsmen need scarcely be surprised at its ability to cross the Channel.

Before the end of September it has commenced to migrate southwards on its way to its winter quarters in Algeria, Egypt, Asia Minor, and Palestine. It is said to be rare in Portugal, and seen on passage only in Spain, touching also at the Azores. It goes, however, much further south, as will be seen presently. Signor Bettoni includes it amongst the birds which breed regularly in Lombardy;[2] and Messrs. Elwes and Buckley note it ("Ibis," 1870, p. 333) as found in Epirus and Constantinople. In Corfu it is met with sparingly in April and September, on its spring and autumn migrations.

---

[1] Jones's "Naturalist in the Bermudas," p. 45.
[2] "Storia Naturale degli Uccelli che nidificano in Lombardia," pt. xxxii. t. 91.

A single instance is on record of its having been shot in Oudh;[1] but Mr. Blyth informed me that he knew of no other authority for it as an Indian bird, although he had seen specimens from Afghanistân. South of the Equator the Landrail penetrates to Natal (*cf.* Gurney, "Ibis," 1863, p. 331), and, according to Mr. Layard ("Birds of South Africa," p. 338), a solitary specimen has been procured in Cape Colony.

Mr. Ayres, who has shot it in Natal, writing of its habits ("Ibis," 1863, p. 331), says: "Having been once flushed, it is a difficult matter to put them up a second time out of the long grass; for, besides running with great swiftness, they have a curious method of evading the dogs by leaping with closed wings and compressed feathers over the long grass some three or four yards, and then, running a short distance, they leap again. The scent being thus broken, they generally evade the most keen-scented dogs; and so quickly are these strange

---

[1] "Bengal Sporting Magazine," 1842, p. 870.

leaps made, that it is only by mere chance that the birds are seen." Many an English sportsman can testify to their power of evading good dogs, although they may not quite know how it is done. Nor is this the only way in which the Landrail displays its cunning. If surprised suddenly and caught, it will often feign death, and remain perfectly stiff and motionless for several minutes, to all appearance dead, but in reality only waiting for an opportunity to steal off unobserved. I have known two or three instances in which this *ruse* has been practised with success upon unsuspecting naturalists. Those who intend, therefore, to investigate the subject of hybernation should be on their guard against what at first sight might strike them as an instance of torpidity.

GENERAL OBSERVATIONS.

IN the year 1872, through the medium of the Natural History columns of "The Field," a series of observations were made by naturalists in different parts of England on the subject of "Our Summer Migrants." A form of calendar was distributed and filled up by each according to his opportunities. In this way, by the end of the year six hundred and forty-five separate observations were placed on record, and it devolved upon me to prepare a report from the statistics so furnished. As a good deal of interesting information was thus brought to light, it occurs to me that I may appropriately bring the present volume to a close by extracting so much of the report as relates strictly to the subject matter in hand, and I accordingly do so.

In the calendars returned, some thirty species of summer migratory birds are mentioned with more or less frequency. The majority of the observations upon them have reference, as might be supposed, to the dates of their arrival and departure, or, more correctly speaking, to the dates when they were first heard or seen and last observed. When referring some time previously to the utilization of such observations, it was remarked that upon various points some addition to our knowledge was desirable. Amongst other interesting facts, for example, might be ascertained the precise line of direction in which various species migrate, the causes which necessitate a divergence from this line, the relative proportions in which different species visit us, the causes which influence the abundance or scarcity of a species in particular localities, the result of too great a preponderance of one species over another, whether beneficial or otherwise to man as a cultivator of the soil, the simultaneity or otherwise of departure from this country in autumn, the causes operating to retard such departure, and so forth. All these are matters of interest, especially to those who reside in the country, and have leisure to inquire into the subject. Let us see how far the observations supplied furnish a reply to these inquiries.

Of the thirty species of migrants mentioned, the Swallow, as might be supposed, has attracted the largest share of attention, and in regard to the number of observations made upon it stands at the head of the list with forty-three. The Cuckoo comes next with thirty-eight; the Chiffchaff and Swift follow with thirty and thirty respectively; and so on through the list, as given below, to the Reed Warbler, upon which bird, strange to say, no more than three observations were made.

The following list will give some idea of the amount of attention which each bird received.

Swallow	43	Redstart	25
Cuckoo	38	Flycatcher	24
Chiff-chaff	30	Landrail	24
Swift	30	Nightingale	23
Willow Wren	29	Martin	23
Sandmartin	27	Tree Pipit	21
Whitethroat	26	Sedge Warbler	17
Blackcap	25	Yellow Wagtail	16

Wryneck	16	Wood Wren	9
Nightjar	16	Pied Flycatcher	9
Wheatear	15	Red-backed Shrike	9
Whinchat	15	Garden Warbler	8
Lesser Whitethroat	14	Reed Warbler	3
Grasshopper Warbler	12	Various	78
Turtle Dove	10		
Common Sandpiper	10		645

The first Swallow was seen, not as might be supposed in the south or south-east of England, but four miles south of Glasgow, on the 2nd of March, and Mr. Robert Gray states that this is the earliest record of its arrival in Scotland. It is, indeed, an exceptionally early arrival, for nearly a month expired before another was seen at Cromer, on the 31st of the same month, and six weeks elapsed between the first and second appearance of the bird in Scotland. On the 1st April, with a S.E. wind, this harbinger of spring arrived at Great Cotes, in Lincolnshire, and on the 3rd of that month was noticed simultaneously at Nottingham and Taunton. From the 6th of April the arrival of Swallows was pretty general until the 13th, when they were first noticed in Ireland at Ballina, co.

Mayo, and on the following day at Glasnevin, Dublin, and at Bray, in the county of Wicklow. The temperature then at Bray was 53°, and the wind S.W. In these localities and dates there is nothing to indicate anything like a precise line of immigration; on the contrary, the birds first appeared where they were least expected. The prevalence of gales, however, at that particular season doubtless operated to retard their progress, and induced them to linger about sheltered localities where food could be obtained. Mr. Wm. Jeffery, who is well situated for observation on the Sussex coast, between the downs and the sea, remarked that most of the spring migrants were several days later than usual in their arrival, and the Swallow in particular not only came later, but lingered longer than is its wont in his neighbourhood. A single bird of this species was seen by him, flying round a steam threshing-machine, on the 10th of December. "Whether it had been disturbed," he says, "from hybernation in the oatrick which was being threshed, or only attracted by the warmth from the engine, I cannot say. It flew very weakly, and was not long seen."

On the 2nd of November, with the temperature at 45·5°, and the wind W., the species was still in the neighbourhood of Huddersfield, and on the 13th November, during cold weather, two were seen on the beach at Exmouth. I may here remark that but little attention is paid to the time of departure of a species compared to that which is given to the date of its arrival.

The Martin was observed to come later and go earlier than the Swallow, the earliest and latest dates being respectively April 10 at Marlborough, and November 7 at Leiston, Suffolk. And in the case of this bird the movement northwards might be traced by the dates, as Wiltshire, April 10; Worcester, April 11; Yorkshire, April 11 and 13 (the weather fine, with temperature 53°, and wind W.); Derbyshire, April 15. Further to the westward, viz., at Llandderfel, in Merionethshire, its appearance was not noticed until the 13th of

May, when the temperature stood at 48°, and the wind N.W. Strange to say, it was not observed in 1872 by any correspondent in Scotland and Ireland.

The Sandmartin is always amongst the first comers to arrive, and was seen in three different counties during the last week of March, viz., in Sussex, Wilts, and Worcester, the weather dull, with the wind blowing from the westward. Its stay in this country is never so prolonged as that of the Swallow, or even the Martin. Large flights are observed preparing to migrate at the end of August and beginning of September, and at the end of the latter month all have gone southward again for the winter. In 1872, however, the species was seen exceptionally as late as October 7.

The Swift is rarely seen before the first week of May or after the first week of August, and of thirty independent observations upon this bird, three only refer to its appearance during the last days of April, four-and-twenty record its arrival between the 1st and 17th of May, and three only relate to its disappearance—from Garvoch, Perthshire, on July 29, from Leicester on August 2, and from Exeter on August 12. It was first seen upon the Devonshire coast at Plymouth and Torquay, and at the former place was particularly numerous. It may be worth noting that a male Swift shot at Cromer by Mr. J. H. Gurney, jun., on June 15, was found to have the under parts denuded of feathers, which would indicate that the males take their turn at incubation.

The Swallows and Swifts are thus brought together, out of the order of the above list, to admit of a more easy comparison of the dates of arrival and departure. We will now follow the order above indicated, commenting only on such facts as appear noteworthy.

The thirty observations which relate to the Cuckoo tend to show that the usual time of its arrival in this country is between the 20th and 27th of April, and in no instance was it observed before the 6th of April (at Torquay) which was considered an exceptionally early date to meet with it. On the Lincolnshire shore it arrived with a southerly wind, in Merionethshire with a west wind, and on the Irish coast with a southwest wind, the weather warm and mild, the temperature 49° to 50·5°.

The most northerly point of observation was Dundee, where it was seen on April 29, but it had been previously noticed at Garvoch, Perthshire, on the 27th, and near Falkirk in Stirlingshire on the 25th of the same month. In Fife, Forfar, and Tayside, Mr. P. Henderson, from personal observation, has frequently found Cuckoos' breasts bare of feathers, as if from incubation, and has observed old birds feeding their own young—a fact in the economy of this bird which has frequently been disputed.

As early as the 2nd of March the Chiff-chaff arrived at Torquay; and, being seen at Chudleigh and Taunton on the 9th, at Northrepps, Norfolk, on the 13th, Hovingham, near York, on the 14th, and Melbourne, Derbyshire, on the 28th, it is easy to trace the gradual movement from south to north of this restless but hardy little bird. A south or south-west wind seems to be most favourable to its arrival, but in this case, as in the case of other species, the data are not sufficient to enable one to judge of this with certainty. It was last seen on Sept. 12 at Sparham in Norfolk.

The Willow Wren was noticed in the midland and northern counties long before its arrival was recorded on the south coast. In Devonshire and Sussex it was observed during the first week of April on various dates from the 3rd to the 7th; in Surrey, Berks, Herts, Norfolk, Lincoln, and Yorkshire somewhat later, that is to say, between the 7th and the 10th of the month; and yet at Nottingham and Melbourne in Derbyshire it was seen upon the exceptionally early date of March 29. In every case where the wind was noted at the time, it was blowing from the W. or S.W., generally from the latter quarter.

Only one notice was supplied of its occurrence in Wales, namely, in the parish of Llandderfel on April 28; but this date does not throw much light upon the progress of the bird westward, for its arrival had already been noted at Glasnevin, co. Dublin, on the 10th, and at Ballina, co. Mayo, on the 13th of the same month. On the last-mentioned date its appearance in Scotland was chronicled in the county of Stirling, but no information was given during that year of its having been observed further north.

In the case of the Common Whitethroat something like a line of migration is indicated by the dates at which the bird was observed. Thus, arriving on the Devonshire, Sussex, and Kentish shores on April 11, 13, and 14 respectively, it was in Berkshire, at East and West Woodhay, on the 15th and 16th; in Leicestershire on the 18th, at Nottingham on the 21st, at Great Cotes in Lincolnshire on the 22nd, at Hovingham, near York, on the 23rd, and by May 6 was as far north as Falkirk. The wind, in all cases where it was noticed, was blowing from the west or south-west, the temperature gradually rising from 48° to 62°.

Of the five-and-twenty observations made

upon the Blackcap, the majority relate to its appearance in the second week of April, and it would seem that in the case of this species, the further north we go, the later the date of its arrival. At Torquay it was observed on the 7th, Marlborough on the 10th, East and West Woodhay, Berks, on the 15th, Barnsley on the 16th, Burton on the 23rd, and Melbourne, Derbyshire, on the 27th. No record was furnished of its occurrence either in Scotland or in Ireland, where it is at all times a rare bird. It was last seen at Nottingham on Nov. 4. The Blackcap, however, does not invariably quit this country in autumn; many instances of individuals having been seen here in mid-winter have been reported by competent observers. It has occasionally happened, however, that the Coal Titmouse (*Parus ater*), which is a resident species, has been mistaken for this bird.

In the West of England, during the year referred to, the Redstart seems to have made its appearance somewhat earlier than usual, having been noted at Bishop's Lydeard, near Taunton, on the 3rd of April. On the 6th it was seen at Keswick, in Norfolk, and on the 8th and three following days in four different localities in Yorkshire; the wind S. W., and the temperature about 51°. After the middle of the month this bird became more numerous, and was very generally observed. In Derbyshire, at Melbourne, it was not seen until April 24, where it seems to have arrived with a S.E. wind; and going still further north, we find it in Stirlingshire and Sutherlandshire on the 27th and 28th. In Ireland it is very rare, and no note was forwarded of its occurrence there in 1872.

The Spotted Flycatcher is always a late comer, seldom appearing before the first or second week in May. Last year, however, it arrived somewhat earlier than usual, and was noticed in Norfolk, at North Runcton, on April 23, and at Barnsley on the 27th; the wind W., and the temperature about 54°, with a haze and light rain. Mr. A. D. Campbell states that Flycatchers were unusually numerous at Garvoch, in Perthshire, about the 21st of May, and were first seen there on the 19th. By Aug. 27 they had all disappeared. Only one note was received of its appearance in Ireland—viz., on May 31, at Ballina, co. Mayo. Mr. Thomas Ruddy, of Palé Gardens, Llandderfel, Merionethshire, referring to this species, says that he saw the old birds in July catching bees, not only in the air, but on the hive-board.

The Landrail, or Corncrake, as it is indifferently called,[1] arrived pretty generally during the last week of April, and was noticed by a great many observers on the 25th, 26th, 27th, and 28th of that month. On the last-mentioned date it was observed in the county of Dublin, and on May 1st at Ballina, co. Mayo. Apparently it did not reach Scotland until a week later, for the first record of its appearance there is on May 8, at Fife. On May 14 and 15 its presence in Stirlingshire and Sutherlandshire

---

[1] Out of twenty-four correspondents, thirteen call this bird the Landrail and eleven the Corncrake, and this in various parts of the country, so that neither name can be regarded by any means as local.

was attested by two good observers. The scarcity of this bird in some seasons is a theme with which readers of "The Field" of late years have become tolerably familiar; but no light is thrown upon the subject, nor is any cause suggested by those from whom calendars were received.

That far-famed songster, the Nightingale, whose notes are so eagerly listened for in early spring, was not heard last year before April 9; but, from causes already referred to, the first utterance of song does not always indicate the earliest arrival, and it is probable that the birds had already been some days in their favourite haunts before their welcome notes betrayed them.

No more favourable locality for this species could be found, perhaps, than that wherein they were soonest observed—namely, the neighbourhood of Ratham, in Sussex. Situated on the flat country between the downs and the sea, about three miles from the former and seven from the latter, with an arm of harbour within two miles, it offers, with its attractions of wood and water, a tempting resting-place to these small winged invaders on their arrival, and furnishes, moreover, a fine post of observation to the inquiring naturalist. Here, throughout April and May, the woods of West Ashling and the copses around Kingley Vale resound with the songs of various warblers, but especially of Nightingales, which find in this safe retreat an immunity from traps which is not everywhere accorded them. On April 10 their remarkable note was detected at Reigate; on the 12th they were singing at East and West Woodhay, in Berkshire; while from the last-named date until April 18 they were daily noticed in various parts of Norfolk and Suffolk. From thence, through Leicester, Derby, and Nottingham, we trace this bird to Yorkshire, where on May 5 we find it at Barnsley, the temperature, according to that good observer Mr. Lister, standing at 50°, and the wind W. Further north than this in 1872 there were no tidings of it, although in former years I have both seen and heard it in the woods by the waterside at Walton Hall, near Wakefield, and have been informed of its occurrence five miles to the northward of York.

I had proposed in these "General Observations" to confine attention strictly to the facts disclosed by "The Field Calendar;" but the subject of the distribution of the Nightingale in England is of such interest to ornithologists, and even to those who, without professing to be naturalists, take a pleasure in listening to the bird, and are not unwilling to learn something about it, that I venture to give an extract from another source which I feel assured will be considered most instructive.

Writing upon this subject in his new edition of Yarrell's "British Birds," now in course of publication, Prof. Newton says (vol. i. p. 315): "In England the Nightingale's western limit seems to be formed by the valley of the Exe, though once, and once only, Montagu, on this point an unerring witness, heard it singing on the 4th of May, 1806, near Kingsbridge in

South Devon, and it is said to have been heard at Teignmouth, as well as in the north of the same county at Barnstaple. But even in the east of Devon it is local and rare, as it also is in the north of Somerset, though plentiful in other parts of the latter county. Crossing the Bristol Channel, it is said to be not uncommon at times near Cowbridge in Glamorganshire. Dr. Bree states ('Zoologist,' p. 1211) that it is found plentifully on the banks of the Wye near Tintern; and thence there is more or less good evidence of its occurrence in Herefordshire, Salop, Staffordshire, Derbyshire, and in Yorkshire to about five miles north of its chief city, but, as Mr. T. Allis states, not further. Along the line thus sketched out, and immediately to the east and south of it, the appearance of the Nightingale, even if regular, is in most cases rare, and the bird local; but further away from the boundary it occurs yearly with great regularity in every county, and in some places is very numerous. Mr. More states that it is thought to have once bred in Sunderland, and it is said to have been once heard in Westmoreland, and also, in the summer of 1808, near Carlisle; but these assertions must be looked upon with great suspicion, particularly the last, which rests on anonymous authority only. Still more open to doubt are the statements of the Nightingale's occurrence in Scotland, such as Mr. Duncan's (not on his own evidence, be it remarked), published by Macgillivray ('British Birds,' ii. p. 334), respecting a pair believed to have visited Calder Wood in Mid Lothian in 1826; or Mr. Turnbull's ('Birds of East Lothian,' p. 39) of its being heard near Dalmeny Park in the same county in June, 1839. In Ireland there is no trace of this species."

It has long been well known that the male birds arrive in this country many days before the females; but, of twenty-three observations made upon the Nightingale, not one refers to or confirms this fact.

The Nightingale has been pictured by poets and naturalists in various romantic situations, but perhaps never before in so unromantic a spot as "under a bathing-machine"! Yet Mr. Monk states that on the 13th of April, 1872, there were "Nightingales on the beach *under the bathing-machines* along the whole length of the shore at Brighton." The explanation which suggests itself is that the birds had just arrived, and had sought the first shelter which offered— a woody shelter, it is true, and a shady one, although of a very different kind to that which the birds had been accustomed to.

The observations made upon the Tree Pipit, twenty-one in number, call for no particular comment, save that the direction of the wind at several dates of supposed arrival was from a S.W. or S.E. quarter, corresponding with what has been observed of other migratory birds, and tending to show that they prefer to travel with a side wind rather than with a head wind or the reverse.

In the eastern counties, for example, it was observed that the Tree Pipit arrived in Norfolk with a S.S.E. wind, the temperature being 52°; in Lincolnshire with a wind veering from

S. to S.E. and E., the weather dry, cold, and
cloudy ; in Yorkshire with a S.W. wind,
weather fine, and temperature 47·5°. It was
first observed at Bushey, in Hertfordshire, as if
arriving directly from the eastward, on the 8th
of April; and was last heard at Ratham, near
the coast of Sussex, on Sept. 15. The furthest
point north at which it was noted was near
Stirling on May 1. In Ireland it is unknown.

In the case of the Sedge Warbler, we again
remark observations on the wind at the presumed
dates of arrival in all respects confirmatory of
what has been already stated. Four good observers in the counties of Norfolk, Lincoln,
Derby, and York noted the direction of the wind
when first meeting with this bird as S.S.E., S.W.,
S.E., and S.S.W., respectively. No record of its
occurrence in 1872 either in Scotland or Ireland
was received. The general period of its arrival
in England seems to be during the last fortnight
of April.

About the same period arrives the Yellow
Wagtail, or Ray's Wagtail, as it is called by

many, respecting which bird sixteen observations were received from different parts of the country. It does not appear to have been met with further north than Wakefield, and no notice was taken of it by correspondents in Scotland and Ireland.

The Wryneck, or Cuckoo's-mate, long preceded the Cuckoo in the date of its arrival, having been heard at Reigate as early as March 31, and at Ratham, in Sussex, on April 2. On the 6th and 7th of the latter month it was observed at several localities in Norfolk, and its appearance generally throughout England in 1872 seems to have been noted during the first fortnight of April. Mr. Lister states that, although found in the neighbourhood of Barnsley in previous years, it was not observed there in 1872.

The Nightjar seems to have been generally met with throughout the country as far west as Llandderfel, in Merionethshire, and as far north as Garvoch, in Perthshire. Mr. Gatcombe observed it in the neighbourhood of Plymouth on April 10, but this must be regarded as an exceptionally early date, for the majority of my correspondents did not meet with it until quite the end of April and beginning of May. On the 15th of June Mr. P. Henderson found two young Nightjars on Tents Muir, Fife, *amidst a colony of terns* (!), and kept them alive for some time on moths, worms, and pieces of raw meat.

The Wheatear and Whinchat received an equal share of attention in the fifteen observations upon each which were forwarded. The first-named appeared at Plymouth as early as March 6, but the observer in this instance, Mr. Gatcombe, states that he hardly ever knew it so early before. It was observed, however, on the same day at Feltwell, Norfolk, by Mr. Upcher; and Mr. Rope reports that in 1871 he saw it at Leiston, in Suffolk, on March 2. In 1872 in the same neighbourhood it did not arrive until March 18, and was much scarcer than in former years. The calendars enable one to trace it that year as far north as Falkirk, where it was seen on April 1; but this is by no means its northern limit, as there is abundant evidence to show.

The Whinchat is not generally seen in this country until the last week of April, and this is confirmed by the notes before me. Mr. J. J. Briggs, however, met with it near Melbourne, in Derbyshire, on April 3; but he appends the remark that he considers this an unusually early date. Mr. J. A. Harvie Browne states that the Whinchat during mild winters occasionally remains in Stirlingshire.

The Lesser Whitethroat was noticed almost exclusively in the midland counties, the earliest date for its arrival being April 12, at Sparham, Norfolk, and the most northerly locality Barnsley. It goes much further north, however, than this, but is considered rare in Scotland, and is unknown in Ireland.

The Grasshopper Warbler was met with throughout the month of April in about a dozen different localities, and, like the last-named species, chiefly in the midland counties. It goes at least as far north, however, as Oban, in Argyleshire. To the westward, it was noted at Taunton in the middle of May. It is a regular summer migrant to Ireland, although in 1872 it was not noticed there by any correspondent.

Like several of the preceding, the Turtle Dove is oftener observed in the southern and midland counties of England, although stragglers are occasionally met with as far north as Northumberland, and even in Scotland. In the Hebrides specimens have been shot in Islay and Skye, but not in the outer islands. Dr. Saxby has recorded several instances of its occurrence in Shetland, and it has twice been procured in Orkney. In Ireland it is regarded as an annual summer visitant to the cultivated districts.

The Wood Wren was noticed nowhere earlier than the 23rd of April, on which date it was heard by Mr. Inchbald at Hovingham, near York; and the paucity of observations on this and the four following species show that they must be very local in their distribution, or less frequently seen than many of their more obtrusive congeners. The Wood Wren apparently comes very much later than either the Chiffchaff or the Willow Wren.

Nine observations only on the Pied Flycatcher were forwarded. These, however, contain one or two notes of interest. The bird has become much commoner of late years, or more observed; and in 1872 it appears to have been met with much further north than usual. A specimen was shot at N. Berwick by Mr. W. Patterson, and exhibited at the Glasgow Natural History Society on the 24th of September, 1872; and another was procured at Biora, in Sutherland, on the 31st of May, by Mr. T. E. Buckley. In Yorkshire it seems to have been very numerous, a score being heard at once in one locality, near York, on the 29th of May. It was found nesting in Norfolk, at Sparham, eggs being laid and the hen bird sitting, on the 3rd of June. To the westward, it was noted at Cirencester; and was found nesting, as in previous years, at Llandderfel, in Merionethshire.

The Red-backed Shrike, or Butcher-bird, is almost confined to the southern midland counties of England, and although stragglers have been met with occasionally in Scotland, it is always regarded as a rare bird there; and in Ireland it is quite unknown. Mr. Cordeaux states that he has never observed it in Lincolnshire. It is always a late comer, seldom, if ever, arriving before the first week in May; and the earliest date recorded for its appearance in any of the calendars is May 2, on which day it was seen at Ratham, near Chichester. Mr. Donald Mathews has observed, in the neighbourhood of Redditch, that it commences nidification immediately on its arrival. The custom which now prevails of "plashing," or laying the tall hedgerows in which the Butcher-bird delights to dwell, has caused it in many localities to forsake haunts where once it was quite numerous. This has been particularly remarked in Middlesex and the counties adjoining.

The observations upon the Garden Warbler, of which eight only are furnished, do not call for any particular comment, save an expression of surprise that a bird with so good a song should not have attracted more attention. The 21st of April is the earliest date recorded for its arrival, at Burton-on-Trent. One would certainly have expected also to find more notice taken of the Reed Warbler, a noisy little bird, whose incessant babbling by reedy ponds and at the riverside makes it almost impossible to overlook it. Nevertheless, but three notes were forwarded of its occurrence in 1872—two in Norfolk, at Lynn and Hempstead, and one in Wilts, at Marlborough; at the last-named place on the 31st of May, at least six weeks after its usual time for arriving. It is not easy to account for its being so overlooked, for it cannot be regarded as by any means a rare bird, although it may be a local one.

Colonel Irby, who has had opportunities of seeing many of our summer migratory birds on passage, from two good posts of observation,

Gibraltar and Tangier, thus refers to the subject in his recently-published volume on the "Ornithology of the Straits of Gibraltar:"—

"Most of the land birds pass by day, usually crossing the Straits in the morning. The waders are, as a rule, not seen on passage; so it may be concluded they pass by night, although I have occasionally observed Peewits, Golden Plover, Terns and Gulls, passing by day.

"The autumnal or return migration is less conspicuous than the vernal; and whether the passage is performed by night, or whether the birds return by some other route, or whether they pass straight on, not lingering by the way as in spring, is an open question; but during the autumn months passed by me at Gibraltar, I failed to notice the passage as in spring, though more than once during the month of August, which I spent at Gibraltar, myself and others distinctly heard Bee-eaters passing south at night, and so conclude other birds may do the same.

"The best site for watching the departure of the vernal migration is at Tangier, where just outside the town the well-known plain called the Marshan, a high piece of ground that in England would be called a common, seems to be the starting-point of half the small birds that visit Europe.

"Both the vernal and autumnal migrations are generally executed during an easterly wind, or Levanter. At one time I thought that this was essential to the passage; but it appears not to be the case, as whether it be an east or west wind, if it be the time for migration, birds will pass, though they linger longer on the African coast before starting if the wind be westerly; and all the very large flights of *Raptores* (Kites, Neophrons, Honey-Buzzards, &c.), which I have seen, passed with a Levanter. After observing the passage for five springs, I am unable to come to any decided opinion; the truth being, that as an east wind is the prevalent one, the idea has been started that migration always takes place during that wind.

Nevertheless, it is an undoubted fact, that during the autumnal or southern migration of the Quail in September, they collect in vast numbers on the European side, if there be a west wind, and seem not to be able to pass until it changes to the east; this is so much the case that, if the wind keeps in that quarter during the migration, none hardly are to be seen.

"On some occasions the passage of the larger birds of prey is a most wonderful sight; but of all the remarkable flights of any single species, that of the common Crane has been the most noteworthy that has come under my own observation.

"On the Andalusian side the number of birds seen even by the ordinary traveller appears strikingly large; this being, no doubt, in a great measure caused by the quantity which are, for ten months, at least, out of the year, more or less on migration; that is to say, with the exception of June and July, there is no month in which the passage of birds is not noticeable,

June being the only one in which there may be said to be absolutely no migration, as during the month of July Cuckoos and some Bee-eaters return to the south."

CONCLUSION.

AS the Swallows are amongst the first to arrive, so they are amongst the last to depart. Long before chill winds and falling leaves have ushered in the month of October, the Warblers, Pipits, and Flycatchers have left the woods and fields, and hurried down to the coast on their southward route. But the Swallows, loth to leave us, linger on far into the autumn, and only bid us adieu when they miss the genial influence of the sun's rays, and can no longer find a sufficient supply of food. The sportsman who crosses the country with dog and gun in October cannot fail to remark the absence of the numerous small birds which were so conspicuous throughout the summer.

The Wheatear has deserted the rabbit warren; the Stonechat and Whinchat have left the furzy common, to make way for the Linnet and the Brambling. In the turnip fields, Thrushes and Meadow Pipits have usurped the place of Whitethroats and Yellow Wagtails; while in the thick hedgerows and coverts noisy Tits now occupy the boughs which were so lately tenanted by the less attractive but more tuneful Willow Wrens.

To the reflecting naturalist, this curious change of bird life furnishes a subject for meditation in many a day's walk, and is a source of much pleasant occupation. Whether we study the birds themselves in their proper haunts, ascertain the nature of their food and their consequent value to man as a cultivator of the soil; or inquire into the cause of their migration, and their distribution in other parts of the world, we have at all times an interesting theme to dwell upon.

From a perusal of the foregoing chapters it will be seen that "our summer migrants" may be classified into certain well-defined groups, according to their structure and habits, and the haunts which they frequent. Upon the wild open wastes and commons we find the Chats, to which family belong the Whinchat, the Stonechat, and the well-known Wheatear. In the hedgerows and copses are to be seen the three species of Willow Warblers—the Wood Wren, Willow Wren, and Chiff-chaff. Wooded gardens and fruit trees attract the Garden Warbler, Blackcap, and Whitethroats; and the thick sedge and waving flags by the waterside shelter the various species of River Warblers. In the open meadows and moist places by the river bank or sea coast we need not search long to find the Pipits and Wagtails; and while the Flycatchers perch familiarly on our garden walls, or pick the aphis off the fruit trees, the Swallows build under our very eaves, and claim our protection for their young. High above all, the noisy Swift holds his rapid, wondrous flight, wheeling and screaming to his heart's content.

CONCLUSION.

At all these birds we have now taken a peep. We have found them in their proper haunts, examined their skill as architects, and their powers as musicians. We have inquired into the nature of their food, the number and colour of their eggs, and their mode of rearing their young; any peculiar adaptation of structure to habits or curious mode of living has been duly noted; and, not content with studying them at home, we have followed these delicate visitors to foreign climes, and found them in their winter quarters.

It is hoped that the reader ere he closes this volume will have gleaned some little information that may be new to him concerning these most interesting families of small birds, whose fairy forms in summer time flit so continually before us, and whose presence or absence makes so great a difference to the naturalist in his enjoyment of a country walk.

INDEX.

BLACKCAP, page 44, 310.
Butcher-bird, 276, 325.
Chiff-chaff, 28, 307.
  ,, Yellow-billed, 29, 30.
Corncrake, 288, 312.
Cuckoo, 219, 306.
Cuckoo's-mate, 242, 320.
Dove, Turtle, 282, 323.
Flycatcher, Pied, 160, 323.
  ,, Red-breasted, 168.
  ,, Red-eyed, 169.
  ,, Spotted, 155, 311.
Goatsucker, 204, 320.
Golden Oriole, 262.
Hoopoe, 249.
Landrail, 288, 312.
Martin, House, 184, 304.
  ,, Purple, 190.
  ,, Sand, 41, 43, 187, 305.
Nightingale, 32, 313—318.
Nightjar, 204, 320.
Oriole, Golden, 262.
Pipit, Meadow, 124.
  ,, Pennsylvanian, 149.

Pipit, Red-throated, page 152.
  ,, Richard's, 142.
  ,, Rock, 130.
  ,, Tawny, 146.
  ,, Tree, 135, 318.
  ,, Water, 138.
Rail, Land, 288.
Red-backed Shrike, 276.
Redstart, Common, 74, 310.
  ,, Black, 78.
Reed Warbler, 101.
Shrike, Red-backed, 276, 325.
Stonechat, 13.
Swallow, 42, 43, 170, 302.
Swift, Alpine, 199.
  ,, Common, 191, 305.
  ,, Spine-tailed, 203.
Turtle Dove, 282.
Wagtail, Grey, 112.
  ,, Grey-headed, 121.
  ,, Pied, 106.
  ,, Ray's or Yellow, 117, 319.
  ,, White, 110.
Warbler, Aquatic, 91.

INDEX.

Warbler, Blackcap, 44.
  " Garden, 59, 326.
  " Grasshopper, 86, 322.
  " Great Reed, 101.
  " Icterine, 29, 30.
  " Marsh, 92.
  " Melodious, 29, 30.
  " Orphean, 51.
  " Reed, 82, 85, 101, 326.
  " Rufous, 103.
  " Savi's, 88.

Warbler, Sedge, 81—85, 319.
  " Willow, 24.
  " Wood, 16, 323.
Wheatear, 1, 321.
Whinchat, 9, 321, 322.
Whitethroat, Common, 67, 309.
  " Lesser, 71, 322.
Willow Wren, 24, 308.
Woodchat, 281.
Wood Wren, 16, 323.
Wryneck, 242.

CHISWICK PRESS:—PRINTED BY WHITTINGHAM AND WILKINS,
TOOKS COURT, CHANCERY LANE.

Now ready; with Illustrations by THOS. BEWICK.   *Demy*
8*vo., cloth gilt, price* 10*s.* 6*d.*

THE NATURAL HISTORY

AND

ANTIQUITIES

OF

SELBORNE,

IN THE COUNTY OF SOUTHAMPTON.

BY THE REV. GILBERT WHITE, M.A.

THE STANDARD EDITION BY E. T. BENNETT.

Thoroughly revised, with additional Notes,

BY JAMES EDMUND HARTING, F.L.S., F.Z.S.

AUTHOR OF "A HANDBOOK OF BRITISH BIRDS," "THE
ORNITHOLOGY OF SHAKESPEARE," ETC.

ILLUSTRATED WITH ENGRAVINGS BY THOMAS BEWICK,
HARVEY, AND OTHERS.

⁂

LONDON:
BICKERS AND SON, 1, LEICESTER SQUARE.
1875.

in my outlet; but were frighted and persecuted by idle boys, who would never let them be at rest.[1]

Three gros-beaks (*Loxia coccothraustes*)[2] appeared some years ago in my fields, in the winter; one of which I shot:

THE HOOPOE.

since that, now and then, one is occasionally seen in the same dead season.

---

[1] The hoopoe is an irregular spring and autumn visitant to this country. It has occasionally nested here, and would do so, no doubt, more frequently if unmolested. Colonel Montagu states, in his "Ornithological Dictionary," that a pair of hoopoes began a nest in Hampshire, but being disturbed forsook it, and went elsewhere; and Dr. Latham, in the Supplement to his "General Synopsis," has referred to a young Hoopoe in nestling plumage, which was shot in this country in May. A pair nested for several years in the grounds of Pennsylvania Castle, Portland (*cf.* Garland, "Naturalist," 1852, p. 82), and according to Mr. Turner, of Sherborne, Dorsetshire, the nest has been taken on three or four occasions by the school-boys from pollard willows on the banks of the river at Lenthay. The birds were known to the boys as "hoops." Mr. Jesse, in a note to this passage in his edition of the present work, states that a pair of hoopoes bred for many years in an old ash tree in the grounds of a lady in Sussex, near Chichester.—ED.

[2] *Coccothraustes vulgaris* of modern systematists.